50 Structure and Bonding

Topics in Inorganic and Physical Chemistry

With Contributions by
M. J. Clarke P. H. Fackler B. M. Hoffman
J. A. Ibers C. K. Jørgensen J. Martinsen
L. J. Pace R. Reisfeld R. J. P. Williams

With 48 Figures and 29 Tables

Springer-Verlag
Berlin Heidelberg GmbH 1982

ISBN 978-3-662-15756-5 ISBN 978-3-540-39082-4 (eBook)
DOI 10.1007/978-3-540-39082-4

Library of Congress Catalog Card Number 67-11280

Originally published by Springer-Verlag Berlin Heidelberg New York in 1982
Softcover reprint of the hardcover 1st edition 1982

2152/3140-543210

Table of Contents

Stacked Metal Complexes: Structures and Properties

James A. Ibers, Laurel J. Pace, Jens Martinsen, and Brian M. Hoffman

Department of Chemistry and Materials Research Center, Northwestern University, Evanston, Illinois 60201, USA

We discuss the structural aspects of crystals that contain molecular stacks and are comprised of two components, at least one of which is a planar transition-metal chelate complex. One of the components may be viewed as an electron donor, D, or its cation, D^{+p}, the other an acceptor, A, or its anion, A^{-q}. In these crystals either one or both components tend to form stacks, and we employ a classification scheme whose first major division is structural: crystals either contain stacks that integrate the D and A units, or the D and A units can be segregated with D or A stacking, or both. The second major division is physical. In an integrated stack crystal the metal-ligand complex is either neutral or ionic; in a segregated stack crystal it either has an integral oxidation state or a non-integral oxidation state. The non-integral oxidation state is sometimes referred to as "partial oxidation", "mixed valence", or "incomplete charge transfer".

The compounds discussed here exhibit a broad range of fascinating properties. Some of these compounds are primarily of structural interest, while others have unusual magnetic properties. Yet others, particularly those involving metallomacrocycles, exhibit high or unusual electrical conductivities.

By classifying and comprehensively reviewing the structural data available, we hope to provide a firmer basis for improving the understanding of the relation between the physical properties and the structural features of these compounds.

Structure and Bonding 50

1 Introduction

In this review we discuss crystals that contain molecular stacks and are comprised of two components, at least one of which is a planar transition-metal chelate complex. One of the components may be viewed as an electron donor, D, or its cation, D^{+p}; the other an acceptor, A or its anion, A^{-q}. Crystals of this type exhibit a broad range of fascinating properties. Some are primarily of structural interest. Others have unusual magnetic properties; for example the first well-documented instance of a spin-Peierls transition was observed in such a material[1]. Yet others, particularly those involving metallomacrocycles, exhibit high or unusual electrical conductivities. It is here that our chief interest lies and here that the gap between the understanding of structures and properties is most apparent. For example, with (phthalocyaninato)nickel iodide, Ni(pc)I[1], a high, metal-like conductivity along the stacking axis is achieved[2,3] whereas bis(diphenylglyoximato)nickel iodide, $Ni(dpg)_2I$[4], has the same structural motif but exhibits very low conductivity.

The total class of stacked donor-acceptor molecular crystals has been extensively reviewed[5–17], but the subclass containing stacked metal chelate and metallomacrocycle complexes is not so well documented. Elsewhere, we have detailed the properties of these systems, integrating the results with structural data when available[18]. Here we focus on the structural aspects of these complexes, although some of their properties are noted. As such, this review attempts to present the structural features of stacked crystals involving metal chelates in a fashion that parallels Herbstein's[15] discussion of organic charge-transfer crystals. We discuss structurally uncharacterized materials only minimally, but include references for completeness. We have also chosen to ignore the ionic platinum systems (e.g., tetracyanoplatinates, oxalates, and squarates), which are amply reviewed elsewhere[6–10].

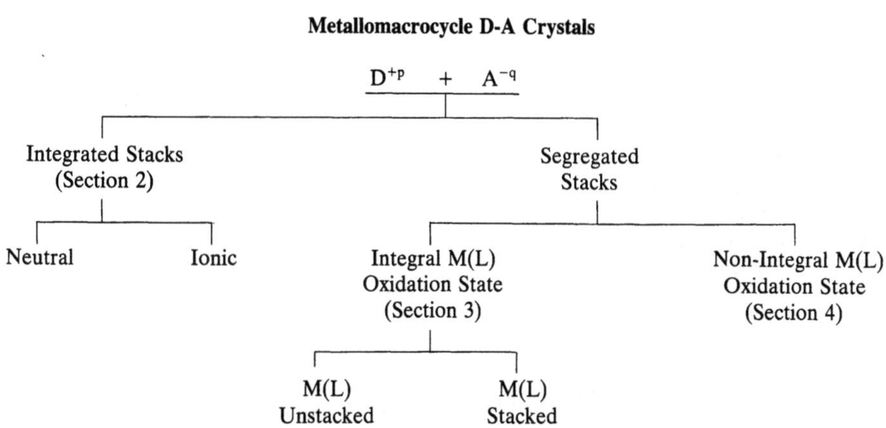

Fig. 1. Scheme for classification of materials prepared from a donor or donor cation, D^{+p}, and acceptor or acceptor cation, A^{-q}, (p, q = 0,1, ...) with one or both components being a planar transition-metal complex

1 Abbreviations for ligands and compounds will be found in Table 1. Table 1 also contains a list of symbols used in ensuing tables.

In bringing together information on what is nonetheless a wide and divergent class of materials, we have found it is useful to employ a variant of a classification scheme for organic donor-acceptor complexes[19, 20]. This is illustrated both by Fig. 1 and by the headings in the Table of Contents. In the crystals discussed here, either one or both components tend to form stacks, and the first major division is structural: crystals either contain stacks that integrate the D and A units, or the D and A units can be segregated with D or A stacking, or both. We define a stack of transition-metal complexes as having parallel units with a metal-metal separation of less than 5 Å and a stack of organic molecules as having an interplanar separation of less than 4 Å. The second major division is physical. In an integrated stack crystal the metal-ligand complex is either neutral or ionic; in a segregated stack crystal it either has an integral oxidation state or a non-integral oxidation state. The non-integral oxidation state is sometimes referred to as "partial oxidation", "mixed valence", or "incomplete charge transfer", and although this state has generally been effected with oxidizing agents, in principle it could equally well be achieved by partial reduction. The classification scheme is convenient for it allows us to group together diverse materials that show similar properties. But it is not unambiguous, as the distinction between integrated and segregated stacks can be arbitrary[21]. Moreover, the many variations in stacking arrangements are not addressed in this classification scheme. To take the single example of the parallel stacking of planar metallomacrocycles, a number of structural motifs are possible (Fig. 2), although the most important by far is the "metal-over-metal" arrangement illustrated in Fig. 2a.

Table 1. Abbreviations

a) Chemical Abbreviations[a]	
aona	N-isopropyl-2-oxy-1-naphthylidenaminato
bipy	2,2′-bipyridyl
bqd	1,2-benzoquinonedioximato
BTF	benzotrifuroxan
CA	chloranil
dhg	1,2-dihydroglyoximato
disn	diiminosuccinonitrilo
dmg	1,2-dimethylglyoximato
dpg	1,2-diphenylglyoximato
dta	dithioaceto
dtc	N,N-di-n-butylthiocarbamato
edt	ethylene-1,2-dithiolato
en	ethylenediamine
etio	2,7,12,17-tetramethyl-3,8,13,18-tetraethylporphyrinato
HMTSF	hexamethylenetetraselenafulvalene
KCP	Krogmann's Salt: $K_2[Pt(CN)_4]Br_{0.3} \cdot 2\,H_2O$
meg	1-methyl-2-ethylglyoximato
mnt	maleonitriledithiolato
NMP^+	N-methylphenazinium
nmsim	bis-N-methylsalicylaldiminato
ntbsim	N-t-butylsalicylideneiminato
oep	2,3,7,8,12,13,17,18-octaethylporphyrinato
omtbp	1,4,5,8,9,12,13,16-octamethyltetrabenzoporphyrinato
ox	oxalato
PA	pycril azide
pc	phthalocyaninato
POZ	phenoxazine
ptt	propene-3-thione-1-thiolato
PTZ	phenothiazine
qnl	8-hydroxyquinolinato
salphen	N,N′-1,2-phenylene-bis-salicylaldiiminato
taab	tetrabenzo[b,f,j,n][1,5,9,13]-tetraazacyclohexadecine
tatbp	triazatetrabenzoporphyrinato
tatma	1,4,8,11-tetraaza-5,7,12,14-tetramethyl[14]annulenato
tbp	tetrabenzoporphyrinato
TCB	tetracyanobenzene
tcdt	1,2,3,4-tetrachlorobenzene-5,6-dithiolato
TCNQ	7,7,8,8-tetracyano-*p*-quinodimethane
tfd	cis-1,2-bis-trifluoromethylethylene-1,2-dithiolato
tmp	5,10,15,20-tetramethylporphyrinato

Table 1 (continued)

TMPD	N,N,N′,N′-tetramethyl-*p*-phenylenediamine
TMTTF	tetramethyltetrathiafulvalene
TNB	1,3,5-trinitrobenzene
tpp	5,10,15,20-tetraphenylporphyrinato
TRNF	2,4,5,7-tetranitro-9-fluorenone
TRP^+	tropylium
TTF	tetrathiafulvalene
TTT	tetrathiatetracene

b) Table Abbreviations

A	acceptor
AF	antiferromagnet
AR	absorption peak reference
C	conductivity reference
CL	Curie law temperature dependence
CWL	Curie-Weiss law temperature dependence
d( )	distances in Å
D	donor
DI	dimer
DC	dichroism
FM	ferromagnet
HAF	Heisenberg antiferromagnet
M	magnetism reference
MM	metal-to-metal
MOM	metal-over-metal stacking arrangement
MT	magnetic transition
NP	not parallel; angle away from parallel follows
PP	plane-to-plane
PS	Pauli susceptibility; nearly temperature independent
R	conventional crystallographie agreement index
S	structure reference
SO	symmetric overlap
SPT	spin-Peierls transition
SS	slipped stack
TAP	temperature activated paramagnetism
TYPE A	metallic conductivity ($\sigma\ \alpha\ T^{-\alpha}$) below room temperature, reaching a broad maximum and then falling off in an activated manner
TYPE B	metallic conductivity ($\sigma\ \alpha\ T^{-\alpha}$) below room temperature, then undergoing a sharp and discontinuous transition to a semiconducting or insulating state
T_m	temperature at which Type A conductivity reaches a maximum
T_{MI}	temperature of metal insulator transition for Type B conductivity
U	uniform stacks
UO	unsymmetric stacks
YM	Young's modulus reference
Δ	activation energy for semiconductor conductivity
σ	conductivity

[a] We have adopted the convention of using lower case letters for ligands and upper case letters for molecules

The classification scheme is particularly effective in arranging crystals by their charge transport properties. Integrated stack crystals (Sect. 2) and integral-oxidation state segregated stack crystals (Sect. 3) are invariably semiconductors, with typical room-temperature conductivities being $\sigma < \sim 10^{-3}\ \Omega^{-1}\ cm^{-1}$. In contrast, many of the non-integral oxidation state, segregated stack crystals (Sect. 4) have high room-temperature conductivities, $\sigma \sim 10 - 500\ \Omega^{-1}\ cm^{-1}$, which increase with decreasing temperature in a metal-like fashion.

Why is this last class of compounds unique in displaying such effective charge transport? It appears that a molecular crystal based on a metal complex (or an organic molecule) can be highly conducting only if it meets two criteria. The first is that the metal-ligand molecules normally must be in crystallographically similar environments and be planar so that they can array in close communication. A crystal satisfying this criterion will exhibit the intermolecular interactions necessary to generate a conductive pathway. However, all of the categories in Fig. 1 potentially may be comprised of crystals containing stacked, planar metal complexes. It appears that a second criterion, that of a non-integral oxidation state, must be satisfied if effective charge transport is to

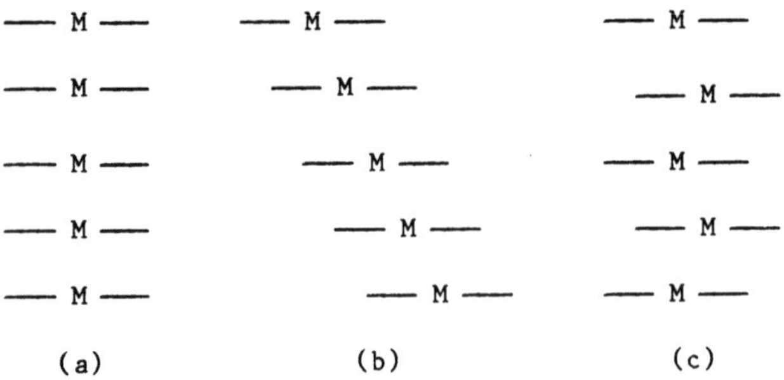

Fig. 2a–c. Some structural motifs for the parallel stacking of planar metallomacrocycles. Motif (**a**), which is the most important, is often called the "metal-over-metal" arrangement. Motif (**b**) is often called the "slipped-stack" arrangement. Many other motifs are possible, (**c**) being but one example

occur[20, 22, 23]. Yet even for a crystal to contain regular stacks of metallomacrocycles in a non-integral oxidation state is apparently not sufficient to ensure high conductivity. Various bisglyoximato complexes appear to possess the necessary requirements of structure and valency, and yet are semiconductors with very modest room-temperature conductivity (Sect. 4). This conundrum remains unresolved.

Consideration of the results collected in this review will make it apparent that chemists recently have developed entirely new classes of materials, composed in part of metallomacrocycles, whose physical properties potentially may be controlled by chemical modification. In some cases these physical properties, for example, metal-like conductivity in the stacking direction, rival those shown by the "organic metals"[6, 8, 11]. We anticipate that as the relation of physical to chemical properties becomes better understood, new tailormade materials with commercially useful physical properties will be synthesized.

2 Integrated Stack Crystals

2.1 General Features

Many compounds exhibit integrated stacks of planar transition-metal complexes and planar π-organics. These systems may be classified either as neutral charge-transfer crystals[24–44] (Tables 2 and 3) or as ionic charge-transfer salts[1, 45–64] (Tables 4 and 5). The nature of the ground state typically has been assigned by an analysis of the optical properties. To a good approximation the optical absorption spectrum of such a compound is given by the sum of the spectra of the component molecules[15, 65] (with an additional band or set of bands, the so-called charge transfer bands, occurring between 5 and 30 kK). The charge of the components typically is assigned by noting whether a better fit of the overall spectrum is obtained using spectra of the neutral donors and acceptors or their ions, but recently other methods have become available[66]. In all known cases, both subunits can be assigned an integral oxidation state (but see below) and the resultant compounds are either insulators or semiconductors.

Table 2. Structure of integrated stack, neutral, 1 : 1 charge-transfer compounds[a]

Compound		Structural data	Ref.
[Ni(tmp)][TCNQ]	$P2_1/n$ Z = 4 R = 0.093	DA(U) along b; UO; d(PP) = 3.43	44
$[Ni(bqd)_2][TCNQ]$	$P\bar{1}$ Z = 1 symmetry $\bar{1}$[b]	DA(U) along c; UO; d(DA) = 3.4	35, 36
$[Pd(bqd)_2][TCNQ]$	$P\bar{1}$ Z = 1 R = 0.064 symmetry $\bar{1}$	DA(U) along c; UO; d(DA) = 3.4	35, 36
$[Pt(ptt)_2][TCNQ]$	$P\bar{1}$ Z = 1 R = 0.037 symmetry $\bar{1}$	DA(U) along a; UO; d(DA) = 3.42	37
$[perylene][Ni(tfd)_2]$	$P\bar{1}$ Z = 1 R = 0.17 symmetry $\bar{1}$	DA(U) along c; UO; angle of DA to c = 32°	33
$[pyrene][Ni(tfd)_2]$		DA(U) along c; SO; small angle to c axis d(PP) = 3.54	33
$[Cu(ntbsim)_2][TNB]$	$P2_1/b$ Z = 4 R = 0.071	DA (pairs) along c; d(DA) = 3.35 $Cu(ntbsim)_2$ non-planar	39
$[Ni(ntbsim)_2][TNB]$	$P2_1/b$ Z = 4 R = 0.091	DA (pairs) along c; d(DA) = 3.35 $Ni(ntbsim)_2$ non-planar	39
$[Co(ntbsim)_2][TNB]$	$P2_1/b$ Z = 4 R = 0.062	DA (pairs) along c; d(DA) = 3.11 $Co(ntbsim)_2$ non-planar	39
$[Pd(qnl)_2][CA]$	$P\bar{1}$ Z = 1 R = 0.107 symmetry $\bar{1}$	DA(NP) along a axis; UO: angle = 15.5° d(Pd-Cl) = 3.44	24, 28, 30
$[Pd(qnl)_2[]TCB]$	$P\bar{1}$ Z = 1 R = 0.102 symmetry $\bar{1}$	DA(U) along a axis; UO; d(PP) = 3.39	24, 29
$[Cu(qnl)_2[]TCNQ]$	$P\bar{1}$ Z = 1 R = 0.195 symmetry $\bar{1}$	DA along a axis; UO; d(PP) = 3.2 d(DA) interplanar = 3.23 d(DD) and d(AA) interplanar = 3.41–3.47	28
$[Cu(qnl)_2][BTF]$	$P2_1/b$ Z = 2 R = 0.195 symmetry $\bar{1}$	DA along a axis; UO	26, 24

[a] All compounds exhibit semiconducting behavior ($\sigma < 10^{-3}\Omega^{-1}cm^{-1}$; activated

[b] Here and in succeeding tables crystallographic symmetry imposed on the metal complex

Table 3. Structure of integrated stack, neutral, 1 : 2 and 2 : 1 charge-transfer compounds[a]

Compound		Structural data	Ref.
$[Cu(salphen)]_2[TCNQ]$	$P\bar{1}$ Z = 1 R = 0.039 symmetry $\bar{1}$	DDA along c; UO; d(DD) = 3.19	40
$[Cu(qnl)_2][PA]_2$	A 2/a Z = 4 symmetry 2	D_A^A along c; UO; d(DA) = 3.4 (average) = 3.1 (shortest)	31 24
$[Cu(qnl)_2][TCB]_2$	$P\bar{1}$ Z = 1 R = 0.082 symmetry $\bar{1}$	D_A^A; UO; d(PP) ~ 3.4 d(DA) = 3.07 (shortest)	32
$[Cu(nmsim)_2][TNB]_2$	$P2_1/c$ Z = 2 R = 0.073 symmetry $\bar{1}$	D_A^A along c; UO; d(PP) ~ 3.4 d(DA) = 3.2 (shortest)	38
$[Co(aona)_2][TCNQ]_2$	$P\bar{1}$ Z = 1 R = 0.037 symmetry $\bar{1}$	D_A^A along c; UO; d(DA) < 3.5	41
$[Ni(etio)][TRNF]_2$	$P2_1/n$ Z = 4 R = 0.078	ADA trimers along b; d(Ni-TRNF) = 3.33	42
$[Pd(dta)_2][Pd_2(dta)_4]$	C2/c Z = 4[b] R = 0.058 symmetry 2 on monomer and dimer	DA along a; d(Pd-Pd) = 2.754 (dimer-dimer) d(Pd-Pd) = 3.399 (monomer-dimer)	43

[a] All compounds exhibit semiconducting behavior ($\sigma < 10^{-3}\Omega^{-1}cm^{-1}$; activated)
[b] The authors incorrectly report a value of Z = 12

Table 4. Structure and properties of integrated stack, ionic, 1 : 1 charge-transfer salts[a]

Compound		Structural data	Properties	Ref.
$[TTF][Cu(tfd)_2]$	$F\bar{1}$ Z = 4 symmetry $\bar{1}$	DA(U) along a, b, and c; d(PP) = 3.90(c); CF_3 rot. disordered; order at 240 °K	S = 1/2 HAF SPT (12 °K)	S 51, 52 M 1, 45, 46 48, 54, 55
$[TTF][Au(tfd)_2]$	$F\bar{1}$ Z = 4 symmetry $\bar{1}$	DA(U) along a, b, and c; d(PP) = 3.93(c); CF_3 rot. disordered; order at 260 °K	S = 1/2 HAF SPT (2.1 °K)	1, 45, 46, 48 54, 55, 72
$[TTF][Pt(tfd)_2]$	$F\bar{1}$ Z = 4 $P\bar{1}$ Z = 1 R = 0.047 symmetry $\bar{1}$	DA(U) along a, b, and c; d(PP) = 3.19(c); CF_3 rot. disordered; order at 270 °K	2 S = 1/2 units FM (300°–40 °K) Intermediate AF ordering at T < 40 °K	S 47 M 45, 50
$[TTF][Ni(tfd)_2]$		DA long a, c; repeat unit along b every 3 chains	2 S = 1/2 units Ferrimagnetic	45, 64

Table 4 (continued)

Compound	Structural data		Properties	Ref.
[POZ][Ni(tfd)$_2$]	$P\bar{1}$ Z = 1 R = 0.050 symmetry $\bar{1}$	DA along b; d(PP) = 3.66	CWL paramagnetism	S 57 C, M 56
[PTZ][Ni(tfd)$_2$]	$P\bar{1}$ Z = 2 R = 0.065	DAADDA along b; d(DA) = 3.36 d(AA) = 3.83 d(DD) = 3.4–3.9	TAP	S 57 C, M 56
[TRP][Ni(tfd)$_2$]	$P2_1/n$ Z = 2 R = 0.137 symmetry $\bar{1}$	DA(NP); UO; angle 17°; d(PP) = 3.9	CWL paramagnetism	63
[Au(dmg)$_2$][AuCl$_2$]	Pnnm Z = 2 symmetry 2/m	DA along c; MOM; d(MM) = 3.26Å	CD Spectra show no M-M interaction	S 60 61, 62

[a] All salts exhibit semiconducting behavior ($\sigma < 10^{-3}\Omega^{-1}cm^{-1}$; activated)

Table 5. Structure of integrated stack, ionic, 2 : 1 and 1 : 2 charge-transfer salts[a]

Compound	Structural data		Ref.
[NMP]$_2$[Ni(mnt)$_2$]	$P2_1/n$ Z = 2 R = 0.041 symmetry $\bar{1}$	DAD trimers along a d(DA) = 3.48 d(DD) = 3.35 cations slightly bent good overlap in trimers	59
[Pt(bipy)$_2$][TCNQ]$_2$[b]	$P\bar{1}$ Z = 1 R = 0.059 symmetry $\bar{1}$	D^A_A along a axis d(DA) = 3.3 two TCNQ moities appear to be σ-bonded	58

[a] All salts exhibit semiconducting behavior ($\sigma < 10^{-3}\Omega^{-1}cm^{-1}$; activated)
[b] Triplet room-temperature EPR spectrum; at 87 °C the paramagnetism increases ($\mu_B = 2.35$; 2 e^- per unit)

Most of the neutral DA crystals have been prepared from an uncharge transition-metal complex and a benzenoid or quinoid organic acceptor. One class of these donors involves the complex [M(qnl)$_2$], where M = Cu or Pd[24–32]. There is one example where a metal-containing subunit acts as the acceptor: M(tfd)$_2$ reacts with the organic donors perylene and pyrene to afford a complex with neutral DA stacks[33, 34].

Compounds of stoichiometry 2 : 1 and 1 : 2 exhibit two different predominant stacking arrangements (Tables 3 and 5). The YXYYXY arrangement, represented by $[NMP]_2[Ni(mnt)_2]$[59], consists of collinear X and Y units that orient plane-to-plane along the stacking axis (Fig. 4). These compounds often show only weak Y-Y interaction (d(Y-Y) > 3.8 Å) in comparison with the X-Y interaction (d(X-Y) < 3.4 Å), and are thus best described as weakly interacting YXY trimers. The X_Y^Y stacking arrangement is common in systems where there is a disparity in the size of the donor and acceptor (*vide infra*).

The strength of the charge-transfer interaction between subunits along a stack depends on the degree of intermolecular π-orbital overlap, and is thus dependent on both the relative orientation of adjacent component molecules within the stack and on the donor-acceptor spacing. The role that charge-transfer stabilization plays in determining the structures of such compounds has been probed theoretically[68]. Calculations indicate that charge-transfer interactions have little effect on the plane-to-plane spacing; rather this distance is probably governed by dipole-dipole and van der Waals' interactions. However, it is found that the interaction energy is very sensitive to the donor-acceptor orientation. Thus, in the absence of sizable orientational contributions from charge-dipole, London dispersion, and hydrogen-bonding interactions, subunits will adopt an orientation that maximizes charge-transfer stabilization. The principle of maximum charge-transfer stabilization is useful in predicting the molecular orientations in many organic DA complexes, but has been less successful in systems with metal-containing subunits because of the presence of large charge-dipole and hydrogen-bonding interactions.

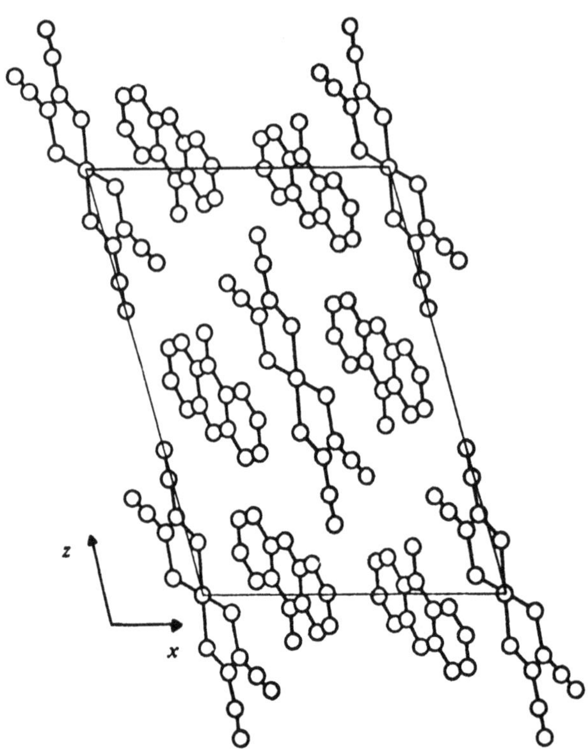

Fig. 4. Projection of the unit cell onto the ac plane showing the DAD trimer stacking arrangement found for $[NMP]_2[Ni(mnt)_2]$ (Ref. 59)

In a majority of these compounds both subunits are closed shell and therefore diamagnetic. Nevertheless, in some a weak intrinsic EPR signal is observed whose intensity is temperature insensitive or decreases with decreasing temperature. Such a signal arises both in complexes with neutral ground states[56)] and in complexes with ionic ground states[1)]. In the former the ground state is diamagnetic, but there is a thermally accessible ionic paramagnetic state. An adequate theoretical treatment of such systems has become available only recently[69, 70)].

2.2 Metal Complex D^{+p} or A^{-q}, Organic Molecule A^{-q} or D^{+p}

2.2.1 Neutral Compounds ($p = q = 0$)

A large number of solid compounds have been prepared from the $M(qnl)_2$ (M = Cu, Pd) donors with acceptors of varying strength and size[24–32)]. Crystal structures, however, have only appeared for a few of these[26, 28–32)]. The compounds of stoichiometry 1 : 1 all stack in

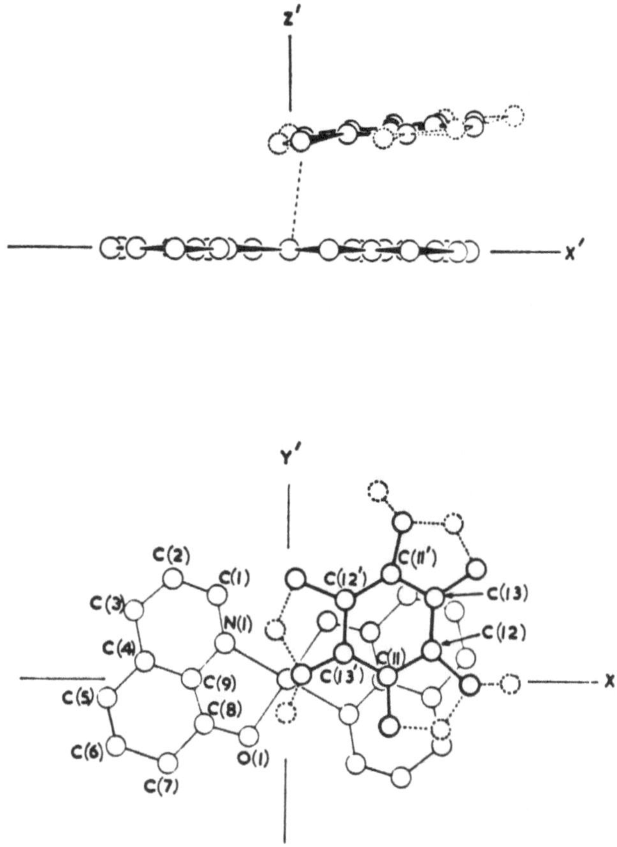

Fig. 5. Projections for $[Cu(qnl)_2][BTF]$ (*top*) parallel to and (*bottom*) perpendicular to the metal complex plane demonstrating the unsymmetrical molecular overlap often found in 1 : 1 $Cu(qnl)_2$ complexes (Ref. 26)

a slipped orientation (Fig. 5), with the acceptors interacting with only the π system of one of the ligands on the donor complex. The slipping is most extreme in $[Cu(qnl)_2][TCNQ]$, where the π systems of the opposite cyano groups interact with two adjacent donor molecules[28]. In spite of the poor direct overlap of metal and acceptor orbitals, the metal still exerts an influence on the charge-transfer interaction, as evidenced by a sensitivity of the optical charge-transfer band energy to the ionization potential of the metal.

Compounds with a 1:2 stoichiometry to date have been obtained only with M = Cu[24, 31, 32]. The units stack in the D_A^A pattern[31, 32] with each A unit interacting with the π system of one of the ligands (Fig. 6). The D and A units are generally not parallel, and have a DA spacing ranging from 3.2 to 3.5 Å. For the systems $[M(qnl)_2][TCB]_x$, crystals have been obtained both for M = Cu[32] and Pd[29]. The former exhibits the 1:2 stoichiometry (x = 2) whereas the latter is 1:1 (x = 1). Murray-Rust and Wright have attributed this change in stoichiometry to the increased stability of an alternative orientation for the first TCB molecule which places it wholly over one half of the Cu complex and leaves a vacancy for the second molecule in a position favorable for charge-transfer interactions[32].

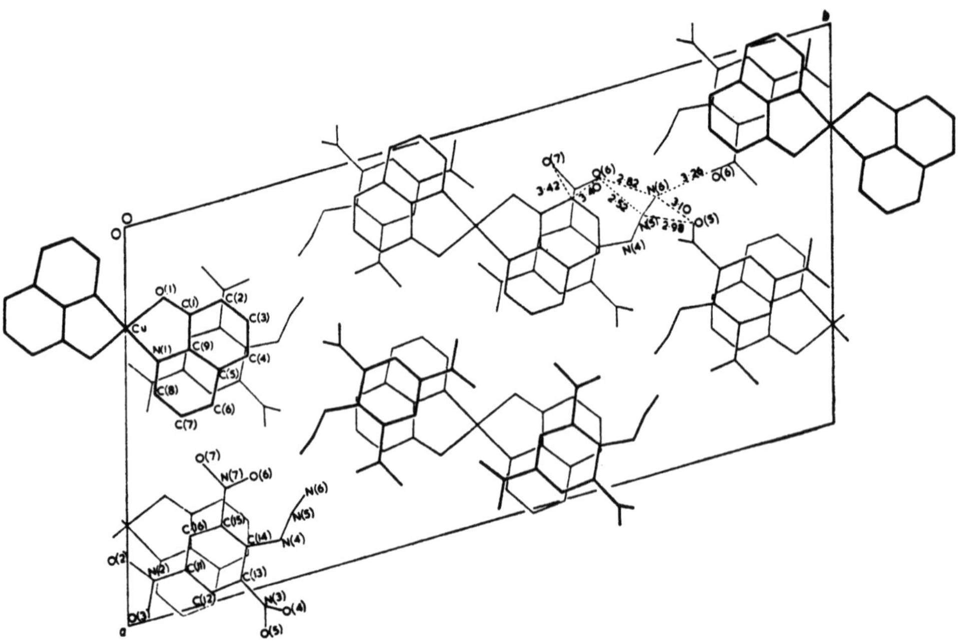

Fig. 6. Projection down the c axis for $[Cu(qnl)_2][PA]_2$ showing picryl azide molecules overlapping with only the organic portion of the metal complex (Ref. 31)

2.2.2 Ionic Compounds (p, q > 0)

Compounds of the type $[TTF][M(tfd)_2]$ (M = Ni, Pt, Cu, Au) exhibit a range of interesting magnetic properties depending on M. In all cases the charge transfer appears complete, and the compounds are thus best formulated as $[TTF]^+[M(tfd)_2]^-$. On the basis of complete structures for the Cu[48, 49] and Pt[47] compounds and precession data for the Au material[48], these three solids appear to be isostructural. The crystals assume a rock-salt packing arrangement, with alternating D and A units along each of the three nearly orthogonal crystal axes in an $F\bar{1}$ cell (Fig. 7). The molecular planes of both subunits lie roughly parallel to the (001) plane, thereby defining the *c* axis as the stacking axis. The intermolecular spacing along this axis is large (~3.9 Å), suggestive of highly localized electronic states; this has been borne out by the low observed electronic conductivity ($\sigma < 10^{-9}\ \Omega^{-1}\ cm^{-1}$)[45]. The crystals may also be viewed as being composed of segregated slipped stacks. By choosing three vectors from the origin of the $F\bar{1}$ cell to the face-centers, one may define a $P\bar{1}$ cell (Fig. 8)[1]. Each of these vectors, labeled a_p, b_p, and c_p, may be viewed as lying along segregated stack axes, although the long interplanar spacings (>7.5 Å) suggest that direct D-D or A-A π-orbital overlap is very small within such stacks.

Each of these compounds is observed to undergo a first-order phase transition at a temperature between room temperature and 200 °K. This transition, which has been fully characterized only for the Cu compound[49], results in the ordering of the rotationally disordered $-CF_3$ substituents, and is probably driven by the accompanying volume decrease of the unit cell. Although the basic site symmetry remains the same as that found at room temperature, there are significant changes in some of the unit cell parameters (Fig. 9). For example, TTF units along the a_p axis move further apart, from 7.80 Å to 8.47 Å, while along the c_p axis the TTF-TTF spacing decreases from 7.53 Å to 6.86 Å. This results in a chain of regularly spaced TTF units along the c_p axis. Preliminary X-ray work indicates that there is a similarly large or perhaps even larger change in the struc-

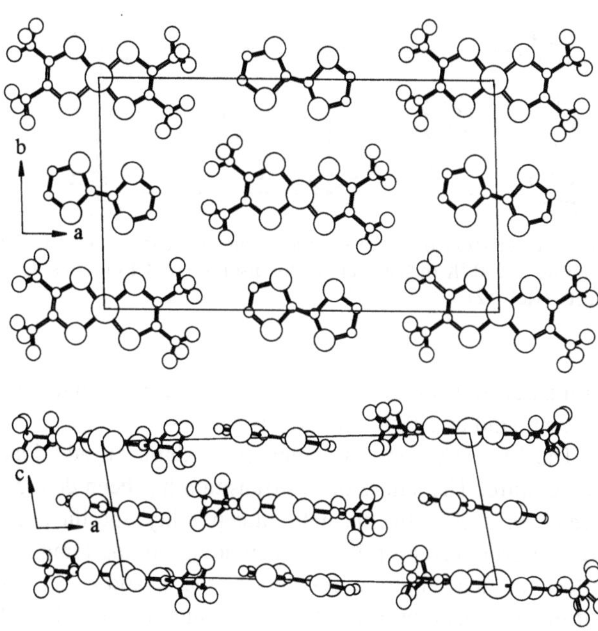

Fig. 7. View of the structure of $[TTF]^+[M(tfd)_2]^-$ (M = Cu, Au, and Pt) for an $F\bar{1}$ cell in the (001) and (010) planes (Ref. 46)

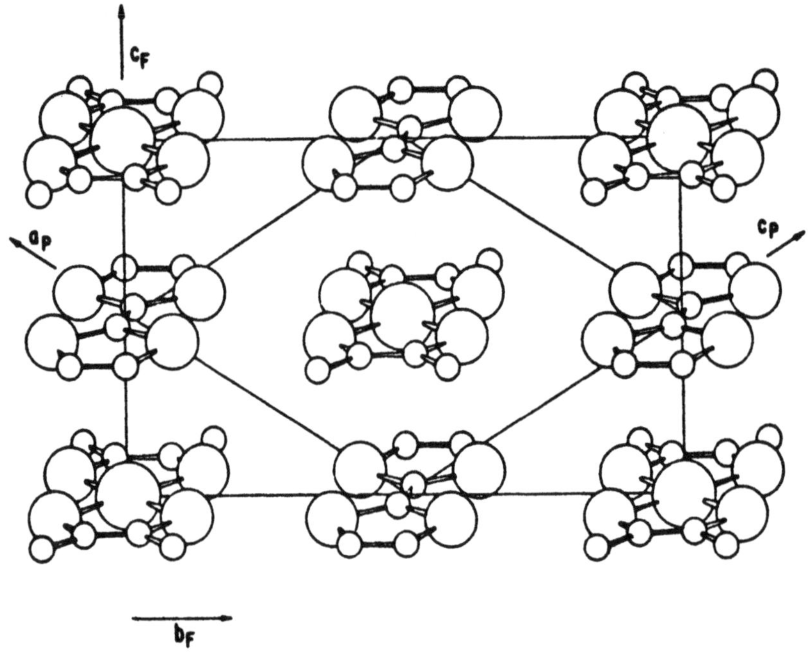

Fig. 8. View of the structure of $[TTF]^+[M(tfd)_2]^-$ (M = Cu, Au, and Pt) in the (100) plane showing the relationship between axes in the $F\bar{1}$ cell and in the $P\bar{1}$ cell (Ref. 1)

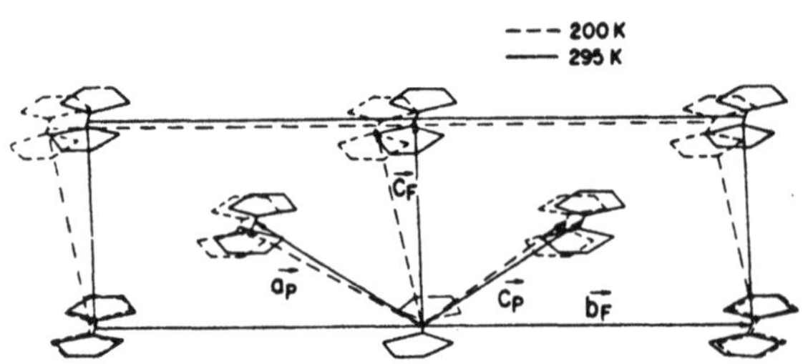

Fig. 9. Illustration of the first order structural change in $[TTF]^+[Cu(tfd)_2]^-$ associated with the ordering at 240 °K of the rotationally disordered $-CF_3$ groups. Only the TTF units are shown for clarity (Ref. 52)

tural parameters of the Au compound at ~200 °K, but that the structural change in the Pt compound at 270 °K is relatively small[1)].

The M = Cu and Au compounds exhibit an additional structural change at low temperature. The nature of this transition has been determined from a variety of measurements, such as the magnetic susceptibility, specific heat, nuclear relaxation, and X-ray scattering taken on the Cu sample at varying temperatures[51-53)]. New superlattice reflections that require a doubling of the a_p and c_p cell dimensions appear below the transition temperature T_c. The TTF units are found to dimerize via translations of 0.036 Å along the c_p axis (Fig. 10), with a net decrease in the TTF-TTF spacing of 0.072 Å. In addition, the TTF units rotate slightly, and there is a movement of the

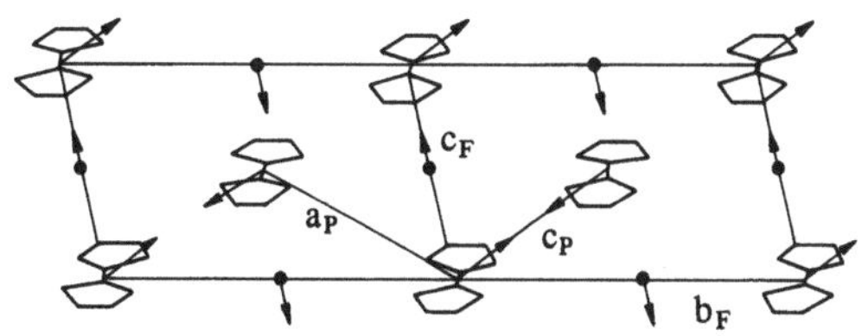

Fig. 10. Illustration of the molecular displacements occurring in the a_p–c_p plane in [TTF] [Cu(tfd)$_2$] below the spin-Peierls transition temperature of 12 °K. Only the [TTF]$^+$ units are shown for clarity; however, the translation of the center of mass of the [Cu(tfd)$_2$]$^-$ units is indicated (Ref. 51)

Cu(tfd)$_2$ units by 0.015 Å in an orthogonal direction to fill in the space vacated by the TTF molecules. This existence of a progressive dimerization in a regular chain of molecules of spin 1/2 along with the observed activated temperature dependence of the magnetic susceptibility below T_c has been interpreted in terms of a spin-Peierls instability[77–78)] in a one-dimensional antiferromagnetically coupled chain.

An additional feature of the temperature dependent X-ray scattering is the persistence above T_c of intensity at the superlattice positions[51)]. This is consistent with a soft phonon mode at a wave vector commensurate with the changes that occur on dimerization. It has been suggested that this low frequency lattice mode may be a requirement for the observation of a spin-Peierls transition.

The systems [PTZ][Ni(tfd)$_2$] and [POZ][Ni(tfd)$_2$] reveal how a seemingly small modification in the donor molecule can cause a dramatic change in solid state properties[56)]. Each compound has been shown by optical spectra to have an ionic ground state. However, in spite of the similarity in the molecular structures and the oxidation potentials of the two donor molecules, the details of the optical and magnetic properties of the two systems differ. The compound [PTZ][Ni(tfd)$_2$] exhibits behavior indicative of strong D^+A^- interactions[79)], whereas [POZ][Ni(tfd)$_2$] shows properties indicative of relatively independent D^+ and A^- π systems. Full crystal structures on both systems[57)] have demonstrated why this is so. The compound [POZ][Ni(tfd)$_2$] stacks in the DADA stacking sequence common for 1 : 1 donor-acceptor complexes, in a nearly eclipsed configuration (Fig. 11, top). The distance between the nickel atom and the least-squares plane of the POZ molecule is 3.66 Å, which is outside the optimum distance for donor-acceptor interactions and therefore accounts for the isolated nature of the D^+A^- π systems. The compound [PTZ][Ni(tfd)$_2$], on the other hand, was the first 1 : 1 compound reported to stack with alternating donor and acceptor pairs (i.e., DDAADDAA) (Fig. 11, bottom). The dominant interactions, however, are not within donors or acceptor pairs (d(DD) = 3.4–3.9 Å, d(AA) = 3.83 Å), but rather between D and A molecules (d(DA) = 3.36 Å). This compound is best described as being composed of interacting DA pairs that are isolated from each other.

The salt prepared from one [Pt(bipy)$_2$]$^{+2}$ and two [TCNQ]$^-$ units has been obtained in two modifications, one paramagnetic and the other nearly diamagnetic at room temperature[58)]. In segregated stacks, TCNQ anion radicals sometimes form spin-paired dimers via π orbital interactions. However, a full structure on the diamagnetic modification of [Pt(bipy)$_2$][TCNQ]$_2$ has revealed that the units stack in a D^A_A arrangement, with no overlap of the π systems on the two TCNQ moieties. Instead it has been proposed that spin pairing occurs via the formation of a long σ bond (1.65 Å) between a quinoid carbon atom on each of the TCNQ units. The quinoid carbon atoms are tetrahedral, with the plane of the carbon atom and its two cyano groups forming an angle of 50° with the plane

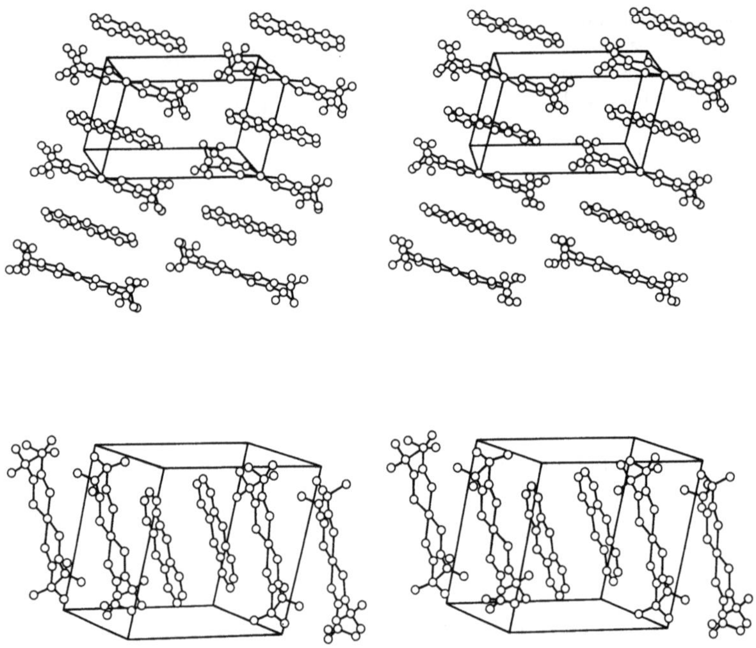

Fig. 11. A stereoscopic view of the unit cell of [POZ][Ni(tfd)$_2$] (*top*) and [PTZ][Ni(tfd)$_2$] (*bottom*) showing the difference in the stacking arrangement for these two compounds (Ref. 57)

of the benzene ring. At 87 °C, a transition occurs that results in a substantial increase in paramagnetism, a color change, and sometimes a macroscopic destruction of the crystals. This has been interpreted as resulting from the breaking of this σ-bond as the TCNQ radicals adopt their planar geometry.

2.3 Mixed Metal Complexes

The complex [Au(dmg)$_2$][AuCl$_2$], one of two systems in Table 3 in which both the D and A units are metal containing, exhibits chains of alternating Au^{III}-Au^{I} ions, with a metal-metal spacing of 3.26 Å[60]. This spacing is similar to the metal-metal distance found in some Ni^{II}, Pd^{II}, and Pt^{II} systems where weak metal-metal interactions are found[80]. However, dichroism[61] and pressure-dependent conductivity studies[82] have shown the absence of any appreciable metal-metal interactions between Au^{I} and Au^{III} ions. The second system involves one of the crystal forms of Pd(dta)$_2$[43]. The structure reveals stacks of alternating Pd(dta)$_2$ monomers and Pd$_2$(dta)$_4$ dimers (Fig. 12). The monomer-dimer metal-metal spacing is 3.399 Å, while the Pd-Pd distance within the dimer is 2.754 Å. The latter distance is shorter than the Pd-Pd distance in Pd(edt)$_2$ (2.79 Å), which has been taken as an example of direct M-M bonding. Moreover, the Pd-Pd distance is 0.14 Å shorter than that between the centers of the S$_4$ planes, another indication of Pd-Pd interactions. Such interactions are consistent with the existence of dimers in solution and

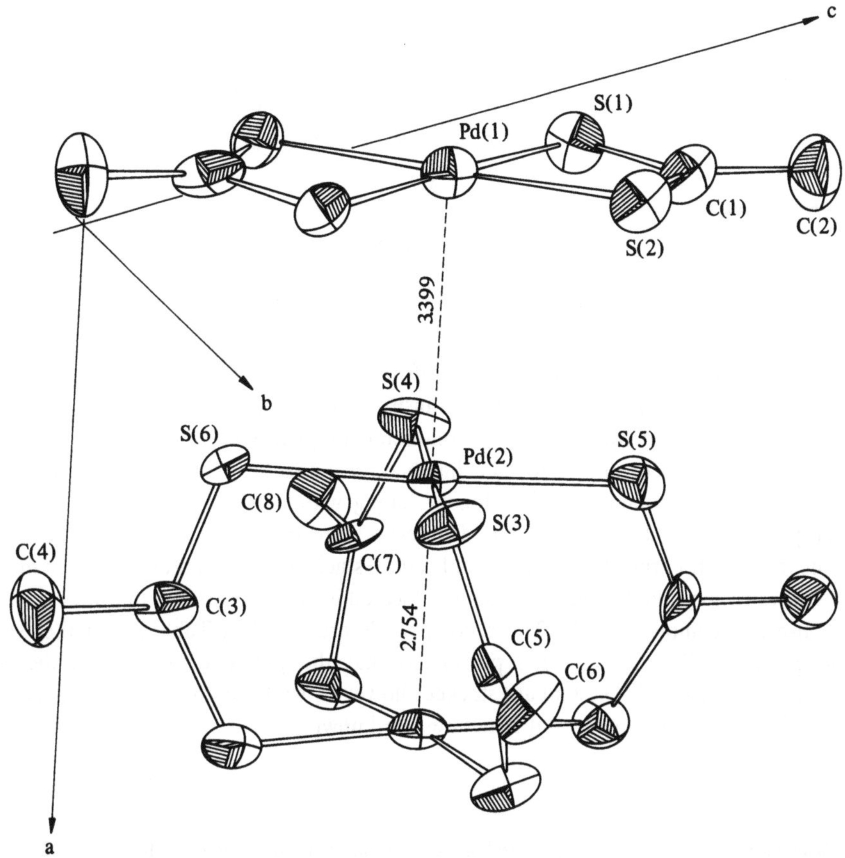

Fig. 12. Illustration of [Pd(dta)][$Pd_2(dta)_4$] unit showing molecular geometry and the Pd-Pd distances (Ref. 43)

in the vapor phase. Single-crystal polarized spectra show properties that can be related to the one-dimensional nature of these interactions, and to the different Pd-Pd separations in a stack.

3 Segregated Stack Crystals, Integral Oxidation State Metal Complexes

3.1 General Features

All crystals in this class are either insulators or semiconductors, and with a single exception (discussed in Sect. 3.3), they are ionic charge-transfer salts. The cations are generally either planar aromatics, such as $TMPD^+$ and TTT^+ (Sect. 3.2) or "onium" counter ions, such as Et_4N^+ and $MePh_3P^+$ (Sect. 3.3), although some cationic metal complexes are also known. For a majority of the systems, the anionic components are bis-1,2-dithiolene metal complexes, which have been studied and reviewed extensively[81–84].

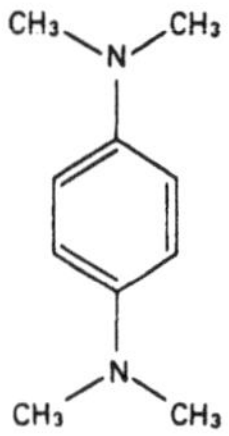

TMPD

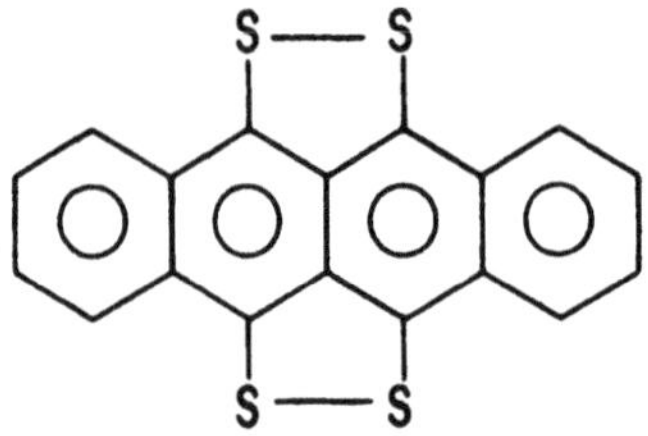

TTT

There are two general structural motifs for segregated stack, integral oxidation state donor-acceptor compounds involving transition-metal complexes. The first, discussed in Sect. 3.2, is one in which the transition-metal complex does not stack and the counter ion, which in all examples to date has been a planar aromatic organic system, does stack[85–91)]. Compounds in this category, along with pertinent structural and physical properties, are included in Table 6. The second type of system, discussed in Sect. 3.3, is one in which the transition-metal complex does stack and the counter ion, which may also be a transition-metal complex, usually is located in columns around the stacks of transition-metal complexes[92–110)]. These compounds are listed in Table 7. On the basis of magnetic studies, it has been postulated that the perylene$^+$ cations in the $[\text{perylene}]_2[M(mnt)_2]$[111–113)] system are stacked; however, we have not included this system in our discussion since no crystal structures are known.

Table 6. Structural data for segregated stack, integral oxidation state crystals with the metal complex not stacked[a]

Compound	Structural data[b]		Ref.
$[TMPD]_2[Ni(mnt)_2]$[c]	$P\bar{1}$ Z = 1 R = 0.042 symmetry $\bar{1}$	stacks of $(TMPD^+)_2$ dimers; d(intra) 3.25 d(inter) 6.57	85
$[Pt(bipy)_2][TCNQ]_3$	$P\bar{1}$ Z = 1 R = 0.073 symmetry $\bar{1}$	stacks of $(TCNQ)_3^{2-}$ trimers; d(intra) 3.23 d(inter) 3.33	86
$[TTF]_2[Ni(edt)_2]$	C2/m Z = 4 R(on F^2) = 0.12 symmetry m	$(TTF^+)_2$ dimers connected by TTF° units; (see text)	S 87 C, M 88
$[TTT]_{1.2}[Ni(edt)_2]$[d]	2 sub-cells $C222_1$ Z = 2 (4.8 TTT) symmetry 2	TTT in uniform stacks;	89, 90, 91

[a] These compounds exhibit semiconducting behavior ($\sigma = 10^{-4} - 30\,\Omega^{-1}cm^{-1}$, activated)
[b] In all cases the transition-metal complex is isolated from the stacks of the organic anion or cation
[c] This compound exhibits a triplet exciton spectrum
[d] AF ($\theta = -5.5$ °K); $Ni(edt)_2$-CWL, TTT-small contribution

Table 7. Structural data for segregated stack, integral oxidation state crystals with the metal complex stacked[a]

Compound	Structural data[b]		Ref.
$[n\text{-}Bu_4N][Cu(mnt)_2]$[c]	I 2/c Z = 8 R = 0.104	DI d(Cu-Cu') = 4.03, 4.43[d] d(Cu-S) = 3.64 (2.17)	S 92, M 93
$[n\text{-}Bu_4N][Fe(mnt)_2]$	I 2/a Z = 8[e] R = 0.074	DI Fe: square pyramidal; d(Fe-Fe') = 3.08, 5.37 d(Fe-S) = 2.46 (2.23)	S 96 M 97, 98
$[Et_4N][Ni(mnt)_2]$	$P2_1/a$ Z = 4 R = 0.050	DI d(Ni-Ni') = 4.14, 4.31 d(Ni-S) = 3.52 (2.15)	S, C 99 C 100 M 93, 97, 101
$[MePh_3P][Ni(mnt)_2]$	$P\bar{1}$ Z = 2 R = 0.090	DI d(Ni-Ni') = 3.47, 3.62 d(Ni-S) = 3.59 (2.15)	S 102 M 97
$[n\text{-}Bu_4N][Co(tcdt)_2]$	$P2_1/c$ Z = 4 R = 0.094	DI (isolated) Co: square pyramidal d(Co-Co') = 3.10, 10.6 d(Co-S) = 2.40 (2.19)	103, 104
$[Ph_3PCl][Au(tfd)_2]$	$P2_1/n$ Z = 4 R = 0.070	DI d(Au-Au') = 4.66, 7.61 d(Au-S) = 3.96 (2.29)	105
$[Pt(disn)_2][PtCl_2(PhCN)_2]$[f]	C 2/m Z = 2 R = 0.029 symmetry 2/m	both species in separate columns held together by H-bonds; d(Pt-Pt') = 4.48	106
$[Ni(tatma)]_2[Ni(edt)_2]$	R = 0.05	stacks of $[Ni(tatma)^+]_2$ dimers separated by $Ni(tatma)^{\circ}$ units, $Ni(edt)_2$ not stacked	107
$[Cu(en)_2][Pt(ox)_2]$	$P\bar{1}$ Z = 2 R = 0.079	zig-zag chains of $Pt(ox)_2^{2-}$; isolated $Cu(en)_2^{2+}$; d(Pt-Pt') = 3.55, 3.86	108
$[Au(dtc)_2][AuBr_2]$	C 2/c Z = 4 R = 0.11 symmetry $\bar{1}$	MOM; adjacent cations staggered d(Au-Au') = 4.06	109
$[Au(dtc)_2]Br$	P 2/c Z = 2 R = 0.09 symmetry $\bar{1}$	MOM adjacent cations eclipsed d(Au-Au') = 4.97	110

[a] These compounds are semiconductors or insulators ($\sigma < 10^{-6}\Omega^{-1}cm^{-1}$)

[b] Distances not supplied in the original paper were calculated using the published atomic coordinates

[c] The EPR spectrum of $[n\text{-}Bu_4N][Ni(mnt)_2]$ diluted in a single crystal of $[n\text{-}Bu_4N][Cu(mnt)_2]$ has been measured (Ref. 94, 95)

[d] M-M' distances are the intra- and inter-dimer distances; MS is the shortest intermolecular MS contact and the average intramolecular MS bond length is given in parentheses

[e] The authors incorrectly reported a value of Z = 16

[f] X-ray photoelectron spectrum: 74.2 eV (Pt^{II})

3.2 Metal Complex Not Stacked

The four systems listed in Table 6 consist of stacks of planar organic molecules surrounded by transition-metal complexes that are not closely associated with each other. All but the $[Pt(bipy)_2][TCNQ]_3$ system contain planar organic cations and metal dithiolene anions. The properties of these compounds vary considerably, so each system will be described briefly.

The first compound in this class to be studied extensively was $[TMPD]_2[Ni(mnt)_2]$[85]. The transition-metal component of $[TMPD]_2[Ni(mnt)_2]$ is diamagnetic and therefore does not contribute to the magnetic properties. The structure consists of stacks of $[TMPD^+]_2$ dimers which are surrounded by isolated $Ni(mnt)_2^{2-}$ anions (Fig. 13). Adjacent $[TMPD^+]_2$ dimers are well-separated with an interplanar distance of 6.57 Å between dimers as compared with 3.25 Å within the dimer. The coupling of spins within the dimer unit results in a singlet ground state with a low lying triplet excited state (a triplet exciton). There is also a very weak coupling interaction between dimers, which allows slow exciton motion along the $[TMPD^+]_2$ chain.

The transition-metal component of $[Pt(bipy)_2][TCNQ]_3$ is also diamagnetic. The structure[86] consists of stacks of trimers, $[TCNQ]_3^{2-}$, surrounded by non-interacting $Pt(bipy)_2^{2+}$ cations. Adjacent trimers are closely spaced with an interplanar separation of only 3.33 Å as compared with 3.23 Å within the trimer. The magnetic properties of this system have been studied much less extensively than those of $[TMPD]_2[Ni(mnt)_2]$. Single crystal EPR spectra do show, however, that spins are interacting along the TCNQ stacks. Structurally, this compound is very different from the 1:2 compound $[Pt(bipy)_2][TCNQ]_2$[58] (Sect. 2) which forms D_A^A stacks with the two TCNQ moieties σ-bonded to each other.

In each of the remaining compounds listed in Table 6, only the transition-metal complex contributes to the magnetic properties. The system $[TTF]_2[Ni(edt)_2]$ has a magnetic moment of 1.67 BM, which corresponds to one unpaired electron per formula unit[88]. On the basis of solid-state optical and EPR spectra, this unpaired electron has been assigned to the $Ni(edt)_2^-$ anion. The absence of any unpaired electrons associated with the TTF^+ cation can be explained by strong electron pairing within $[TTF^+]_2$ dimers. These dimers are located along the *b* axis and are connected along the *b* and *c* axes by neutral TTF molecules[87]. The molecular planes of the $Ni(edt)_2$ units are parallel to the bc plane, as shown in Fig. 14. The TTF units in the dimer are eclipsed with intermolecular

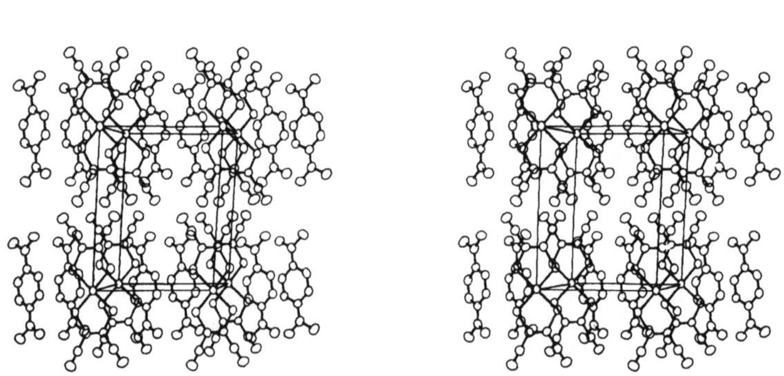

Fig. 13. A stereoscopic view of the unit cell of $[TMPD]_2[Ni(mnt)_2]$ (Ref. 85)

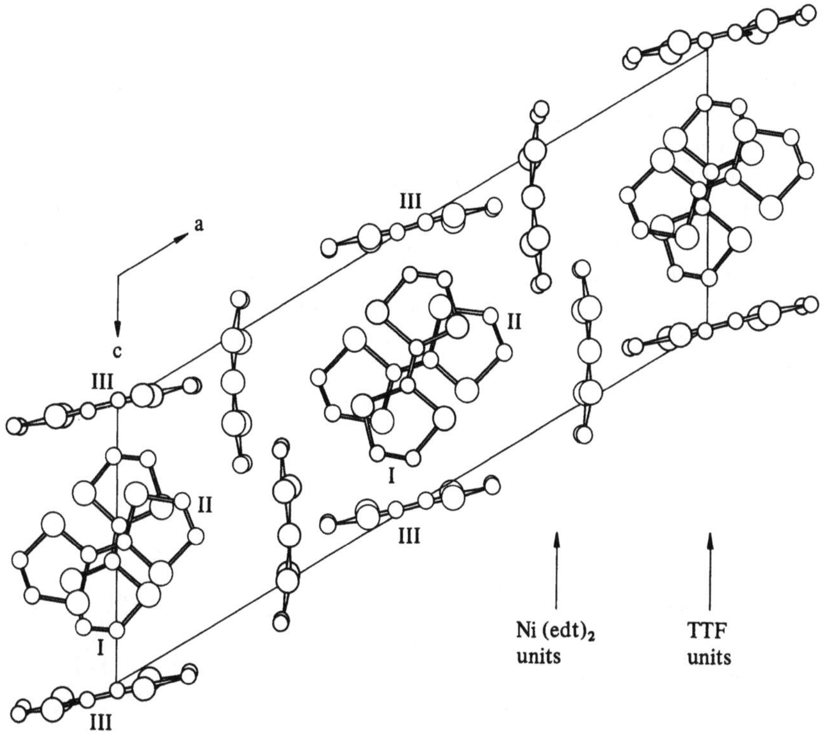

Fig. 14. The unit cell of $[TTF]_2[Ni(edt)_2]$ viewed down [010]. Three different types of TTF units are identified by Roman numerals (Ref. 87)

S-S contacts of 3.48 Å. Given the overall composition of the unit cell, this compound is most instructively formulated as $[TTF^{\circ}]_2[TTF^{+}]_2[Ni(edt)_2^{-}]_2$.

The $[TTT]_{1.2}[Ni(edt)_2]$[89–91] system is similar to the $[TTF]_2[Ni(edt)_2]$ system in that only the unpaired electron from the $Ni(edt)_2^{-}$ anion contributes to the susceptibility. However, here the TTT units are uniformly spaced in the stack, and there are no crystallographically distinguishable TTT° and TTT^{+} entities. This suggests that the electrons are delocalized along the stack; the most accurate formulation for this complex is then $[TTT]_{1.2}^{+5/6}[Ni(edt)_2]^{-}$, with the organic donor ion in a non-integral oxidation state and the transition-metal complex in an integral oxidation state. It is postulated that the contribution to the susceptibility of the unpaired spins on the TTT^{+} cations is quenched by a highly correlated one-dimensional (Peierls) distortion.

Electrical conductivity measurements of the $[TTF]_2[Ni(edt)_2]$ and $[TTT]_{1.2}[Ni(edt)_2]$ systems show them to be semiconductors with conductivities of $\sim 10^{-3}$ and 30 $\Omega^{-1}cm^{-1}$, respectively. In both compounds, the direction of highest conductivity is along the stacks of organic molecules. Since the $[TTT]_{1.2}[Ni(edt)_2]$ system contains chains of partially oxidized organic molecules that appear to be uniformly spaced along the stacking axis, a high conductivity would be expected. The low room-temperature conductivity (compared with other partially oxidized TTT salts) and its temperature dependence, which is typical of a semiconductor, are presumed to be a consequence of the postulated Peierls distortion.

3.3 Metal Complex Stacked

Most of the compounds listed in this class (Table 7) are tetraalkylammonium (R_4N^+) salts of metal dithiolene complexes. These salts generally exist in a 2:1 or 1:1 stoichiometry. The 2:1 salts, such as $[n\text{-}Bu_4N]_2[Cu(mnt)_2]$[114] and $[Et_4N]_2$-$[Cu(mnt)_2]$[115, 116] can exhibit interesting properties, but they will not be considered here since the transition-metal components are isolated from each other (d(M-M) > 5 Å).

All of the 1:1 metal dithiolene systems listed in Table 7 are composed of a phosphonium or ammonium cation and a metal^{III} bis-dithiolene monoanion. These compounds are structurally very similar, consisting of slipped stacks of metal dithiolene anions which are associated in pairs; the stacks are surrounded by non-interacting cations. A view of a representative unit cell for this type of system is shown in Fig. 15. Only $[n\text{-}Bu_4N][Cu(mnt)_2]$[92] displays a slightly different association of anion pairs in the

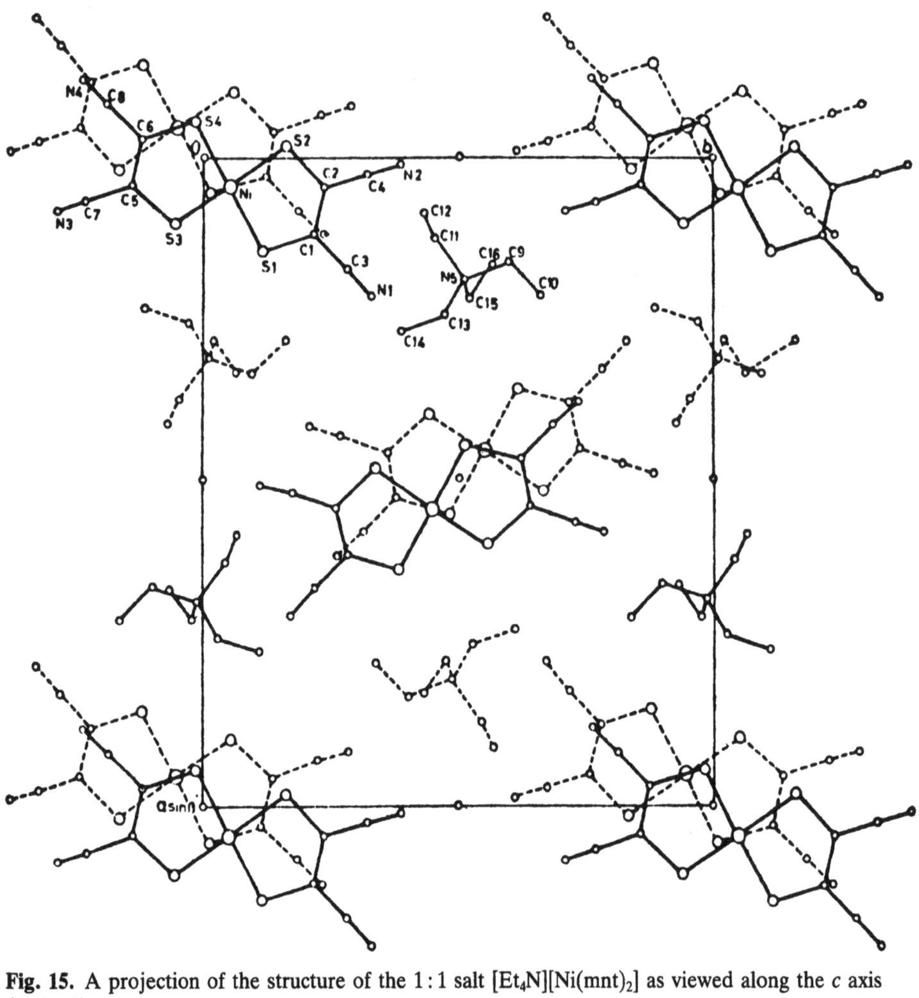

Fig. 15. A projection of the structure of the 1:1 salt $[Et_4N][Ni(mnt)_2]$ as viewed along the c axis (Ref. 99)

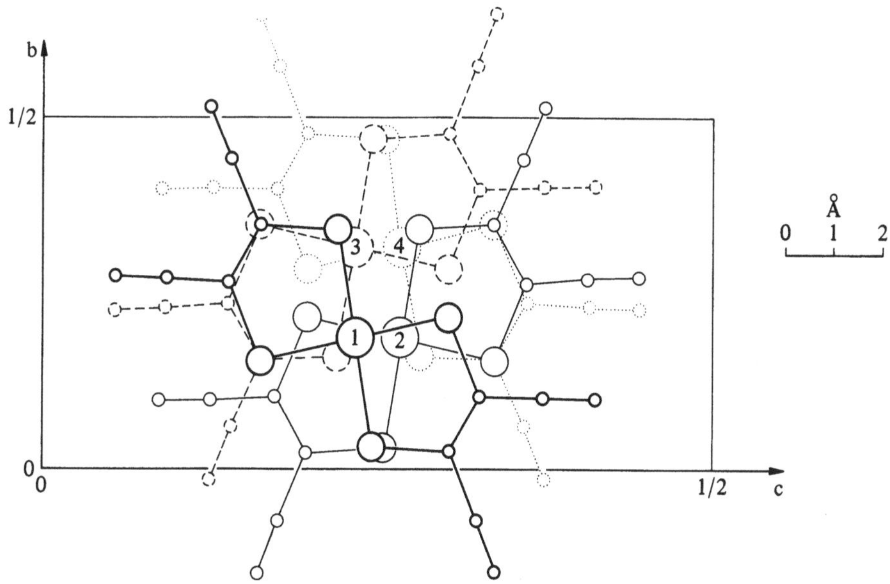

Fig. 16. A projection down the *a* axis of [*n*-Bu_4N][$Cu(mnt)_2$] showing the stacking of the [$Cu(mnt)_2$]$^-$ anions as viewed down a stack (Ref. 92)

stack (Fig. 16). To avoid classifying compounds incorrectly we have included in Table 7 only those compounds for which a full X-ray crystal structure has been reported. The well-known hazard of using X-ray powder patterns to determine whether compounds of similar overall composition have identical crystal structures arose yet again when the original claim[93)] that [*n*-Bu_4N][$Co(mnt)_2$] and its Cu analogue were isomorphous had to be withdrawn on the basis of differences in intensities from single crystals[117)].

As has been noted previously[81)], the structural parameters of metal dithiolene complexes are relatively insensitive to the overall charge on the complex, and the monoanionic dithiolene complexes generally retain the mmm (D_{2h}) molecular symmetry found in both neutral and dianionic dithiolene complexes. The only notable structural trend is a lengthening of the M-S bond as the overall charge on the transition-metal complex increases.

The transition metal in the bis-dithiolene monoanions is formally trivalent. The complexes with Au^{III} and Cu^{III} are diamagnetic, those with Ni^{III}, Pd^{III}, Pt^{III}, and Fe^{III} exhibit a doublet ground state, and that with Co^{III} exhibits singlet-triplet behavior. A reduction in the magnetic moments of the Ni, Pd, and Pt systems, as well as the absence of an EPR signal[118)], has been found in the solid state[93, 101)]. This has been attributed to a pairwise spin-spin interaction where the $S = 1/2$ transition-metal complexes are antiferromagnetically coupled into singlet ground state pairs in the solid. The metal-metal separation in these systems is generally too large to allow direct spin correlation through the metal atoms. Consequently, this correlation is postulated[97)] to occur through an intermediate sulfur atom by way of a M-S---M pathway (Fig. 17) that is accessible because the metal dithiolene anions dimerize by forming two M---S linkages[117)] (Fig. 18). The dimerization can be very weak, as is apparent in the structures of [Et_4N][$Ni(mnt)_2$][99)] and [$MePh_3P$]-

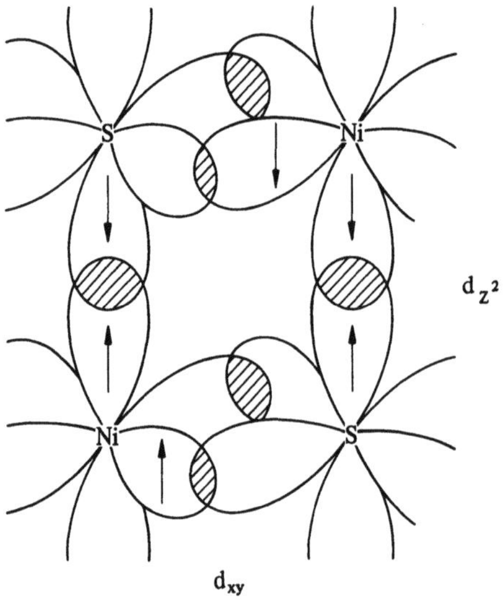

Fig. 17. Schematic representation of proposed spin correlation mechanism in $[Ni(mnt)_2]^-$ systems (Ref. 97)

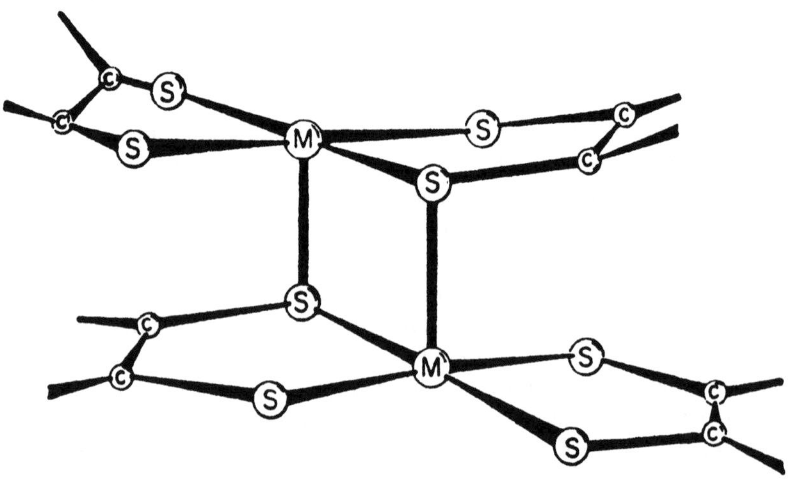

Fig. 18. Idealized structure showing the two M--S linkages that form upon dimerization of two dithiolene complexes (Ref. 117)

$[Ni(mnt)_2]$[102] where there is less than 0.2 Å difference between the distance within and between dimers. This dimerization also can be very strong as in $[n\text{-}Bu_4N][Fe(mnt)_2]$[96], $[n\text{-}Bu_4N][Co(tcdt)_2]$[104], and $[Ph_3PCl][Au(tfd)_2]$[105]; in each of these systems the distance within the dimers is much less than the distance between them. In $[n\text{-}Bu_4N][Co(tcdt)_2]$ the distance between dimers is so large (> 10 Å) that the overall structure resembles that of the 2 : 1 R_4N^+ salts of metal dithiolene systems with the isolated dimer units in the 1 : 1 salts in place of the isolated monomer units in the 2 : 1 salts. In both the Co and Fe systems, the linkage within dimers is so strong that the intermolecular M---S distance is nearly equal to the intramolecular M-S distance, and the transition metal is best described as possessing a square-pyramidal coordination geometry. The effect of dimerization, whether weak or strong, on the metal dithiolene complex is limited to minor deviations of the MS_4 unit from planarity[81]. In the more tightly bound dimers[96, 104], the metal atom is displaced by about 0.3 Å out of the plane of the four sulfur atoms.

Electrical conductivity measurements have been made on a few of these metal dithiolene systems and show that they are, at best, poor semiconductors, with room-temperature conductivities of less than $10^{-6}\Omega^{-1}cm^{-1}$. This is not surprising as the molecules are in an integral oxidation state. Moreover, the existence of dimers destroys the crystallographic uniformity of the stacks and, as the magnetic studies have shown, the unpaired electrons tend to couple within the dimers rather than delocalize along the stacks.

Of the five compounds in Table 7 that do not contain a metal dithiolene system, the most interesting structurally is $[Pt(disn)_2][PtCl_2(PhCN)_2]$[106]. The compound, which is made by cocrystallizing $Pt(disn)_2$ and $PtCl_2(PhCN)_2$, consists of segregated stacks of the component molecules with an interplanar distance of ~3.40 Å. A view of the structure of $[Pt(disn)_2][PtCl_2(PhCN)_2]$ along the stacking (*c*) axis is shown in Fig. 19. Both Pt atoms are still in the Pt^{II} state as measured by x-ray photoelectron spectroscopy. Moreover, it is unlikely that the disn ligand alone has been reduced since $M(disn)_2$ systems have much lower electron affinities than the closely related $M(mnt)_2$ systems[119]. Thus, $[Pt(disn)_2][PtCl_2(PhCN)_2]$ is the only example of a 1:1 DA compound consisting of neutral component molecules that does not crystallize in stacks of alternating donor and acceptor molecules (Sect. 2). Stabilization of the segregated stacking arrangement can be attributed to intermolecular hydrogen bonding between the imine hydrogen atoms and the chlorine atoms on the $PtCl_2(PhCN)_2$ molecule. Although uniform, segregated stacks are present in $[Pt(disn)_2][PtCl_2(PhCN)_2]$, the components are not partially oxidized, and consequently, the electrical conductivity is low ($<10^{-9}\Omega^{-1}cm^{-1}$ for a pressed pellet).

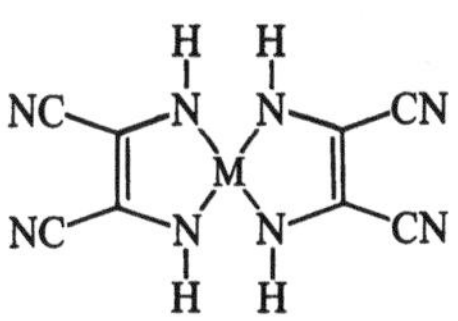

$M(disn)_2$

Another interesting compound in this class is $[Ni(tatma)]_2[Ni(edt)_2]$[107], which is structurally similar to $[TTF]_2[Ni(edt)_2]$[87] (Sect. 3.2) and contains stacks of dimers, $[Ni(tatma)^+]_2$, connected by neutral Ni(tatma) molecules. These stacks are similarly

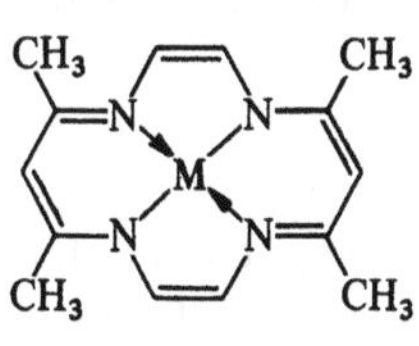

M(tatma)

Fig. 19. A perspective view of $[Pt(disn)_2][PtCl_2(PhCN)_2]$ looking along the *b* axis. The column on the left contains the $PtCl_2(PhCN)_2$ molecules, and the column on the right contains the $Pt(disn)_2$ molecules (Ref. 106)

surrounded by the unstacked, paramagnetic $[Ni(edt)_2]^-$ anions. The compound [Ni(tatma)]$_2$[TCNQ] has also been prepared, but powder diffraction data indicate that it has a different structure.

4 Segregated Stack Crystals, Non-Integral Oxidation State Metal Complexes

As noted in the Introduction, effective charge transport in crystals of metallomacrocycles is limited to this category. With but one exception to date[120, 121], all such metallomacrocycles contain the metal surrounded by an essentially square-planar array of four coordinating nitrogen atoms. The most fruitful approach to the preparation of highly conducting molecular solids based on metallomacrocycle building blocks has used complexes that may be viewed as variants on the metalloporphyrin skeleton[2, 3, 122–138].

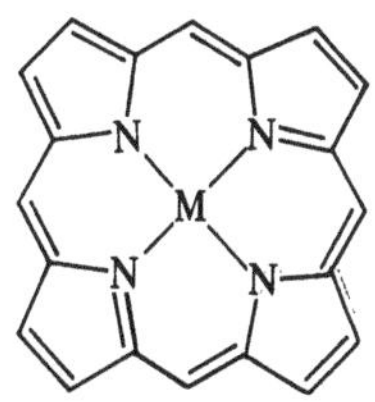

M(porphyrin)

The chemical flexibility of the porphyrin-like metallomacrocycles[139–141] provides an unusual opportunity to vary purposely and rationally the electronic properties of the molecular building block, and therefore of the resulting solid, through choice of the basic ligand, of peripheral substituents, and of incorporated metal. Variants to the porphyrin skeleton[142] have been examined extensively:

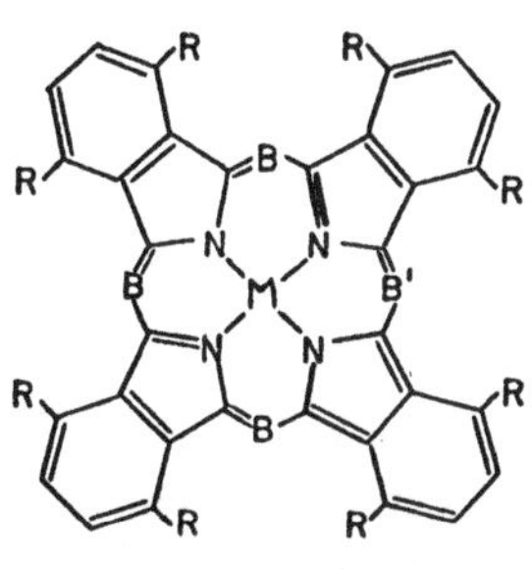

R = H, B=B'=N; M(pc)
B=N B'=CH; M(tatbp)
B=B'=CH; M(tbp)
R = CH_3, B=B'=CH; M(omtbp)

The tetrabenzoporphyrin(tbp) skeleton can be viewed as having an expanded aromatic π-electron system obtained by fusing benzene rings onto the β-carbon atoms of the porphyrin pyrrole rings. Phthalocyanine(pc) may be imagined to arise from four methine → aza bridge substitutions in tbp. In addition, an "intermediate" complex, triazatetrabenzoporphyrin(tatbp), which can be obtained conceptually from tbp by three methine → aza bridge substitutions, has also been studied.

Thus, this class of complexes contains two sizes of ring systems, and within each size the chemical and physical properties can be modulated by choice of bridge atom. Moreover, the basic structures are the foundation for an enormous range of complexes: almost every metal ion in the Periodic Table can be inserted into the central hole of these doughnut-shaped structures, and the periphery of the macrocycles is also susceptible to substitution. Such substitutions, which are well known for the small-ring porphyrin system[139,140], lead, for example, to tetramethylporphyrinato (tmp)[143] and octaethylporphyrinato(oep) derivatives, and the octathioalkyl tetraazaporphyrins[144], and may also be applied to the large ring, tetrabenzporphyrin(tbp) system to give such complexes as M(omtbp) and analogously substituted phthalocyanines.

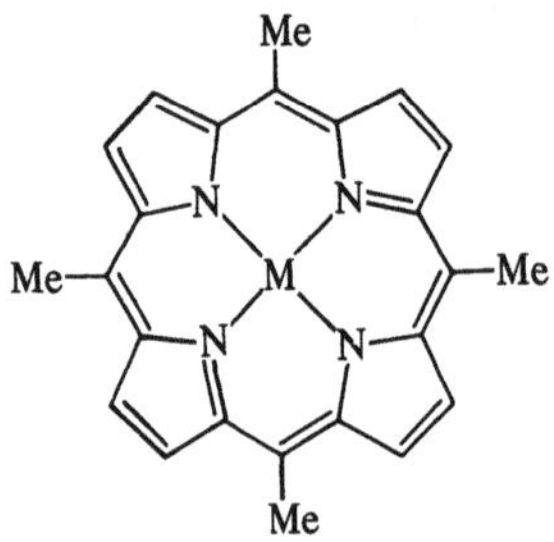

M(tmp)

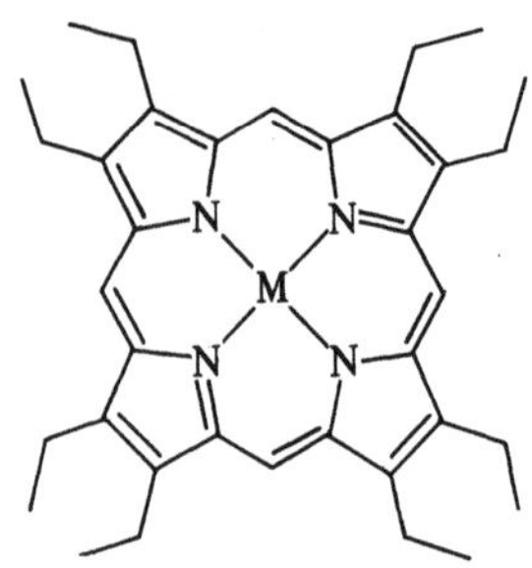

M(oep)

A second series of compounds, based on the bis(α,β-dionedioximates) of various transition metals, especially the Ni triad, has been extensively studied over a longer period[4,145–171].

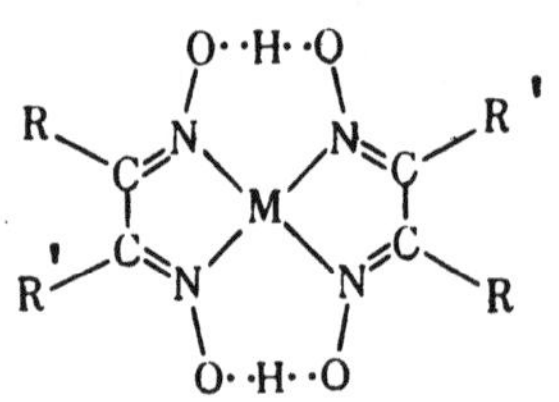

R=R'=H:	$M(dhg)_2$
R=R'=CH_3:	$M(dmg)_2$
R=R'=C_6H_5:	$M(dpg)_2$
R=CH_3, R'=C_2H_5:	$M(meg)_2$

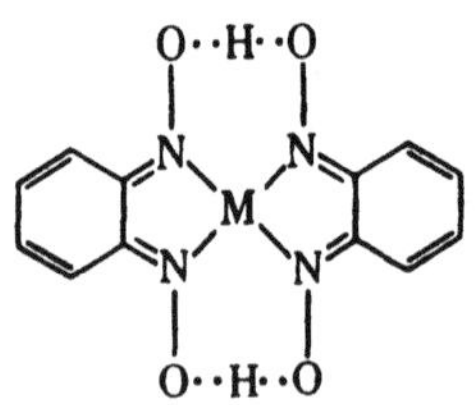

$M(bqd)_2$

This series was the first to be subjected to partial oxidation through the use of iodine[171]. The integral oxidation state parent complexes of Ni, Pd, and Pt are insulators, while the partially oxidized derivatives are semiconductors. These complexes are of particular interest within the context of this review because of the diversity of crystal structures they adopt and because they serve as a contrast to related systems, the prophyrins and phthalocyanines, that display more interesting physical properties. As we have noted, one condition for metal-like conductivity in the stacking direction is a structural motif that involves stacked columns of closely interacting, planar metallomolecules (Fig. 2a). An interesting puzzle is to understand those factors that result in the bis(α,β-dionedioximates) of various metals occasionally adopting this motif, yet failing to exhibit high conductivities.

Finally, the first reports[120,121] of a conducting sulfur chelate, involving a partially oxidized metal dithiolate, have appeared as have reports of materials based on dibenzotetraaza[14]annulenes[172–175], further analogues of the porphyrins.

We first give an overview of the general features of these various compounds and the means by which they have been characterized (Sect. 4.1). The individual classes of compounds are then described, beginning with the dionedioximates (Sect. 4.2), followed by the most important class, the prophyrinic complexes (Sect. 4.3), and then, briefly, by the dithiolates and porphyrin-like analogues (Sect. 4.4).

4.1 General Features and Methods of Characterization

Although the generation of compounds in which the metal component is in a non-integral oxidation state can be accomplished in principle by either reduction or oxidation, all highly conducting crystals reported to date have been prepared by halogen oxidation of the neutral parent molecule. A major reason that partially reduced materials have not been much studied is that, unlike the oxidized materials, they are generally air sensitive. Molecular iodine has been an especially advantageous oxidant because of the stability of polyiodide anions (I_3^-, I_5^-) in nonpolar environments and because such ions are readily accommodated in channels in the various structures. Molecular bromine may be similarly useful, but it has not been employed frequently. Other means of oxidation are, of course,

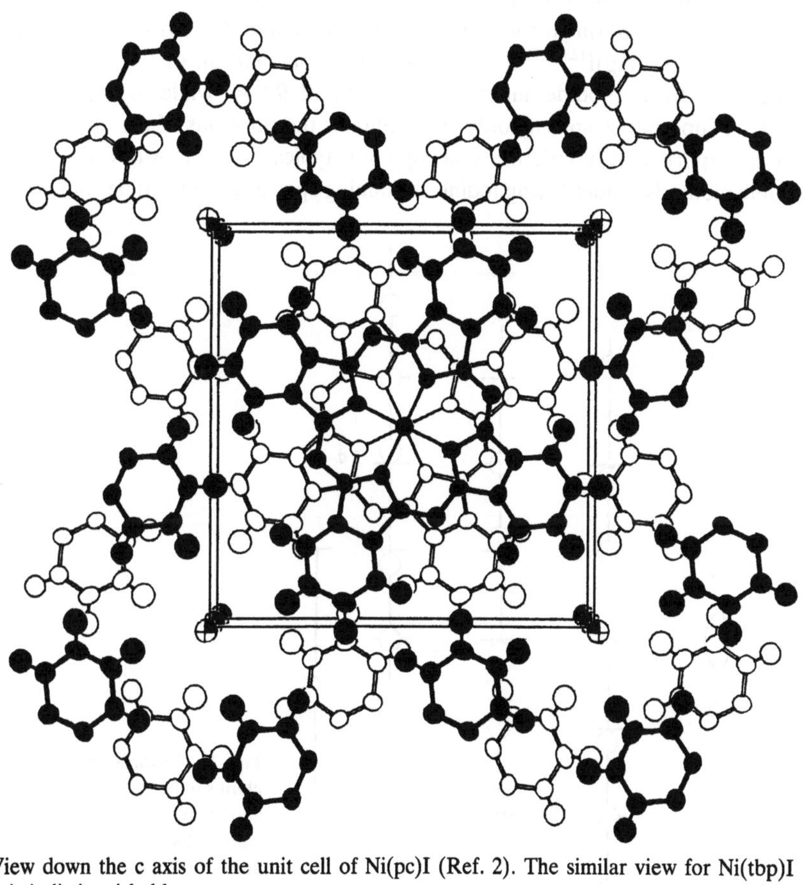

Fig. 20. View down the c axis of the unit cell of Ni(pc)I (Ref. 2). The similar view for Ni(tbp)I (Ref. 132) is indistinguishable

applicable. All of the iodine-oxidized systems discussed here crystallize with stacks of metallomacrocycles around which are four parallel channels containing linear chains of iodine (Fig. 20) with one exception: In the $Pd(bqd)_2I_{0.5} \cdot 1/2\,S$ system (S = aromatic solvent), only two of the channels contain linear chains of iodine while two contain solvent[162]. In all cases metallomacrocycles are stacked metal-over-metal such that the mean molecular plane is perpendicular to the stacking (*c*) axis (Fig. 2a). They alternate in an ABAB pattern along *c*, where molecule B is related to molecule A by symmetry but is rotated about the stacking axis by either ~40° in the porphyrin and phthalocyanine systems, ~65° in the bqd systems, or 90° in the dpg systems.

The iodine species is usually disordered, as indicated by the presence of diffuse X-ray scattering in planes perpendicular to the stacking direction[4,164,176]. In the Ni(pc)I[2], Ni(tbp)I[132], and $Ni(bqd)_2I_{0.5}$[164] systems, it has been possible to index all of the diffuse lines on the basis of a superlattice spacing of 9.5–9.8 Å, which is typical for a triiodide superlattice. The diffuse X-ray intensities for Ni(pc)I and $Ni(bqd)_2I_{0.5}$ have been collected and analyzed, and for each a model consisting of ordered chains of symmetrical triiodide units, which are disordered (translated) with respect to their neighbors, is consistent with the data (Fig. 21). Often the intensity of a given diffuse line is slightly modulated, which indicates that there is a degree of three-dimensional order, but this feature of the diffuse scattering has not been investigated. For several compounds the diffuse X-ray scattering is too complex to be indexed on the basis of a single superlattice spacing. In Ni(tmp)I[128,135] for example, most, but not all, of the diffuse lines can be indexed with a triiodide superlattice spacing of 9.8 Å, while in $Ni(omtbp)I_{1.08}$[125], a 19.46 Å superlattice spacing only approximates the positions of the diffuse lines. There are no reports of a superlattice spacing that is typical of the I_5^- ion. Convenient spectroscopic methods of identifying iodine-containing species have been developed in order to

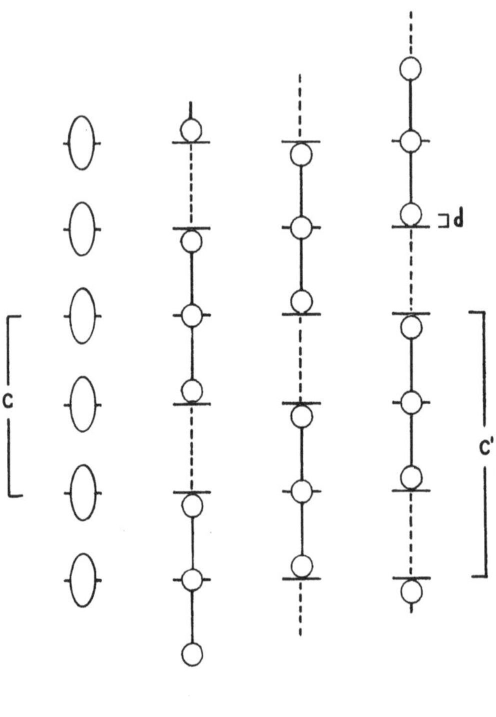

Fig. 21. Disorder model for the iodine chains in Ni(pc)I. Average positions are shown on the left; the three distinct supercells each containing three such sites are shown with the same scale on the right (Ref. 2)

allow rapid characterization of newly prepared $M(L)I_x$ materials[177,178]. Resonance Raman spectroscopy offers a simple means of characterizing polyiodide anions and ^{129}I Mössbauer spectroscopy can be used to quantify the amounts of I^-, as well as to support the Raman assignment of polyiodide species[2].

An understanding of charge transport in a molecular crystal must begin with knowledge of a number of characteristics of the material, such as composition and structure, but knowledge of the ionicity, or degree of partial oxidation, is also essential. In many systems the ionicity is difficult to obtain[6,8,11]. However, the techniques applicable to halogen oxidized donors eliminate such difficulties. A knowledge of the composition of an $M(L)I_x$ crystal and characterization of the anionic species by resonance Raman and Mössbauer spectroscopy and X-ray diffraction techniques makes it possible to determine the ionicity directly. For example, a crystal of composition M(L)I in which the iodine-containing species is shown to be I_5^- ions is immediately characterizable as $(M(L))^{+0.2}(I_5^-)_{0.2}$. However, the use of spectroscopic methods without corroboration can be misleading. Reaction of the dication $[Pd(taab)]^{+2}$ with molecular iodine produces a compound formulated by chemical analysis and resonance Raman spectroscopy as the non-integral oxidation state compound $Pd(taab)^{+2.7}(I_3^-)_{2.7}$[179,180]. However, later the crystal structure showed the compound to be $[Pd(taab)][I_8]$, an integral oxidation state system containing the unusual I_8^{-2} polyiodide dianion[181].

Describing the cation state generated from a neutral $M^{II}(L)$ metallomacrocycle is more complex than describing an oxidized organic compound. One must consider the site of oxidation, since both the central metal and the aromatic π-system of the macrocycle can be redox active[182–184]. For some metals, such as Mn, Fe, Co, and Ag, the first oxidation normally occurs at the metal ion and the process can be written

$$M^{II}(L) \rightarrow M^{III}(L) + e^-$$

However, for metals without a readily accessible trivalent state, such as Mg and Zn, the highest occupied orbital of the complex is a π-molecular orbital of L, and the first oxidation is ligand centered.

$$M^{II}(L) \rightarrow M^{II}(L^+) + e^-$$

Neutral molecules with an anion bound to a trivalent metal also behave in this manner.

There are instances, however, where the metal and ligand oxidized states of an individual complex are roughly degenerate and are interconvertible. The oxidized form of Ni(tpp) has been shown to exhibit an equilibrium between the $Ni^{II}(tpp^+)$ and $Ni^{III}(tpp)$ forms, with the former exhibiting a typically free-radical like EPR signal and the latter displaying an EPR spectrum characteristic of a low-spin, d^7, Ni^{III} ion[184]. Similarly, Ni(pc) can form either a ligand or metal-oxidized species, depending on conditions[182,185]. EPR measurements provide the means of ascertaining the nature of the carriers in a partially oxidized stack of M(L) complexes[125,132], somewhat akin to the procedures utilized in studies of the organic molecular metals[6,11,186]. As we discuss below, borderline cases become particularly interesting when they occur in conductors.

4.2 The α,β-Dionedioximates

The bis(α,β-dionedioximates) of various transition metals, especially dmg, dpg, and bqd complexes of the Ni triad, have been extensively studied[4, 145–171]. Table 8 lists those glyoximate systems for which complete structural data are available. Table 9 lists the corresponding data for the $M(bqd)_2$ systems. These complexes adopt two basic structural motifs. The first, typified by α-$Pd(bqd)_2$[158], consists of planar units stacked in columns with the M-M vector parallel to the stacking axis (Fig. 2a). The repeat in such a motif is usually two units, with a rotation of approximately 45° between successive units. The metal-metal distances are short (3.15 to 3.40 Å), the packing is somewhat inefficient, leading to relatively low density and the existence of channels in the stacking direction, and even the unoxidized parents show evidence of anisotropic properties[157]. The channels serve not only as a potential residence for an oxidant, such as a halogen, but also for solvent molecules (occasionally undetected[187]). Figures 22 and 23 show two views of the $Ni(dpg)_2I$ structures[4]. The channels are clearly evident. But other arrangements of channels are possible, as shown in Fig. 24, where the metal complexes maintain this same motif. The second motif (Fig. 2b), typified by $Ni(bqd)_2$[156] (Fig. 25), involves the placement of planar units in a slipped-stack arrangement so that the M-M vector is not parallel to the stacking direction. The result is a longer M-M distance, a more efficient packing with higher density, no channels, and often a herringbone arrangement. Clearly the tendency to form metal-metal interactions favors the first motif, while intermolecular van der Waals' and packing forces favor the second motif.

In addition, polymeric complexes of $M(dpg)_2$ and $M(dmg)_2$, M = Fe, Co, have been prepared with pyrazine as a bridging ligand. The structure consists of pyrazine-M-pyrazine-M··· infinite chains with dmg or dpg units perpendicular to the chains[188].

Through the synthesis of a large number of bis(α,β-dionedioximates) of the Ni triad Endres et al.[189] have attempted to define those factors that influence the motif. Although they carried out no structure determinations of the resultant compounds, they did determine unit cell dimensions and crystal systems. Based on the correspondence of these cells with those of known structures (Tables 8 and 9) they attempted to deduce the motif present and in the case of motif 1 the M-M distance from the unit cell dimensions. Such an approach involves risks, as similar unit cells do not necessarily guarantee similar structures. They conclude that the first motif, involving direct metal-metal interactions, is favored for the Ni triad when strongly electron withdrawing ligands of low bulk are employed. But probably more subtle factors are involved. Thus, while $Ni(bqd)_2$[156] (Fig. 25) adopts motif 2, $Ni(bqd)_2I_{0.018}$[162] (Fig. 24) adopts motif 1; it is by no means clear that the small amount of iodine present occupies a specific position in the structure. Similarly, $Pd(bqd)_2$ exists in two modifications as does $Ni(meg)_2$. But when $Ni(meg)_2$ is recrystallized from $CHCl_3$, only the modification adopting the second motif is obtained[190]. Clearly the packing forces in these systems are deserving of further theoretical consideration.

The question of structural motif aside, these complexes are of interest owing to the changes in physical properties that occur on partial oxidation. The report by Edelman[171] that $Ni(dpg)_2$ could be oxidized by I_2 or Br_2 to afford $Ni(dpg)_2X$ (X = Br, I) appears to be the first in what is now an extensive group of reports on the oxidation of the bis(α,β-dionedioximates) of the Ni triad as well as of planar metallomacrocycles in general. Edelman logically guessed that $Ni(dpg)_2X$ was a compound of Ni^{III}, a relatively unusual,

Table 8. Structures of metal glyoximates in integral and non-integral oxidation states[a]

Compound	Space group	R Index	Symmetry	Motif	d(M-M) (Å)	d(M-N) (Å)	N-M-N (deg)	Ref.
$Pd(dhg)_2$	$P\bar{1}$	0.046	none	SS	3.558(3)	1.957(7), 1.991(8) 1.953(9), 1.958(7)	80.3(4), 80.3(4)	S 145
$Ni(dhg)_2$	$P2_1/c$	0.039	$\bar{1}$	SS	4.196(1)	1.868(4), 1.880(4)	82.2(2)	S 146, also see 147
$Pt(dhg)_2$	$P2_1/n$	0.043	$\bar{1}$	SS	3.504(1)	2.013(14), 1.968(14)	78.0(7)	S 148
$Ni(dmg)_2$	Ibam	0.12	2/m	MOM	3.245	1.85(2), 1.85(2)		S 149, 150 C 151
$Ni(meg)_2$	$P2_1/c$	0.048	$\bar{1}$	SS	4.747(1)	1.861(4), 1.862(4)	82.5(2)	S 152 (earlier 153)
$Pd(dmg)_2$	Ibam	0.065	2/m	MOM	3.25	1.99(2), 1.93(2)		S 150, C 151
$Pt(dmg)_2$	Ibam	0.07	2/m	MOM	3.23(1)	1.93(4), 1.95(4)	83	S 154, C 151
$Ni(dpg)_2I$	P 4/ncc	0.093	222	MOM (I_5^-) (ruffled)	3.223(2)	1.868(15)	82(1)	S 4, C 4, 169, 170
$Pd(dhg)_2I$	$P4_2/mcm$	0.064	mmm	MOM	3.244(1)	1.989(19)	79.3(10)	S 168
$[Pt(dpg)_2][ClO_4]$	P 4/ncc	0.079	222	MOM (ruffled)	3.259(4)	1.95(2)	77(1)	S 155

[a] The unoxidized parent molecules are insulators $\sigma < 10^{-9}\Omega^{-1}cm^{-1}$ at room temperature; the iodine oxidized materials are semiconductors ($10^{-2} > \sigma > 10^{-5}\Omega^{-1}cm^{-1}$)

Table 9. Structures of metal benzoquinonedioximates in integral and non-integral oxidation states[a]

Compound	Space group	R Index	Symmetry	Motif	d(M-M) (Å)	d(M-N) (Å)	N-M-N (deg)	Ref.
$Ni(bqd)_2$	$P\,2_1/n$	0.062	$\bar{1}$	SS	3.856(3)	1.868(4), 1.860(5)	83.6(2)	S 156, YM 157
α-$Pd(bqd)_2$	Imcb	0.087	2/m	MOM	3.202(1)	2.00(2), 1.95(2)	81(1)	S 158, YM, AR 157
β-$Pd(bqd)_2$	$P\,2_1/c$	0.052	$\bar{1}$	SS	3.774(2)	1.963(8), 1.967(8)	79.5(4)	S 159, YM 157
$Pt(bqd)_2$	Ibam	0.052	2/m	MOM	3.173(1)	1.99(1), 1.99(1)	80.0(7)	S 160, YM, AR 157, C 161
$Ni(bqd)_2I_{0.018}$	Ibam	0.11	2/m	MOM	3.180(2)	1.858(10), 1.940(10)	83.0(5)	S, C 162
$Ni(bqd)_2I_{0.52} \cdot 0.32\ C_6H_5CH_3$	P 4/mcc	0.095	2/m	MOM (I_3^-)	3.153(3)	1.91(2), 1.90(2)	84(1)	S 162, 163, 164, AR 157, C162
$Pd(bqd)_2I_{0.50} \cdot 0.52\ o$-$C_6H_4Cl_2$	P 4/mcc	0.052	2/m	MOM (I_3^-)	3.184(3)	1.996(7), 1.955(7)	80.8(3)	S, C 162, YM, AR 157
$Pt(bqd)_2 \cdot 1/2\ AgClO_4$	P 2/n	0.048	$\bar{1}$	SS	3.386(3)	1.99(2), 1.95(2) 1.98(2), 1.99(2)	77(1) 77(1)	S 165
$Pt(bqd)_2I_2$ Modification I	$P\bar{1}$	0.085	$\bar{1}$	[b]	–	2.00(1), 2.01(1)	80.0(6)	S 166
$Pt(bqd)_2I_2$ Modification II	$P\bar{1}$	0.035	none	[b]	–	2.011(6), 2.024(6) 1.987(6), 1.980(6)	79.4(3), 78.9(3)	S 166

[a] The unoxidized parent molecules except $Pt(bqd)_2$ are insulators; the $I_{0.50}$ substances and $Pt(bqd)_2$ are semiconductors

[b] Packing of discrete, octahedral Pt^{IV} centers

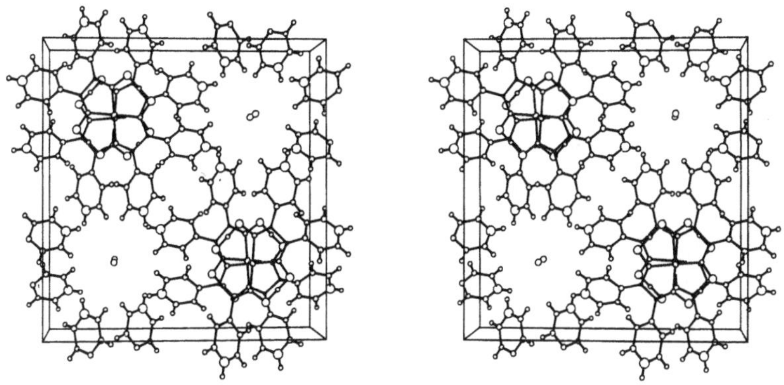

Fig. 22. A stereoview of the unit cell of $Ni(dpg)_2I$. The *a* axis is horizontal to the right, the *b* axis is vertical from bottom to top, and the *c* axis is toward the reader. The vibrational ellipsoids are drawn at the 50% level, except hydrogen atoms, which are drawn arbitrarily small (Ref. 4)

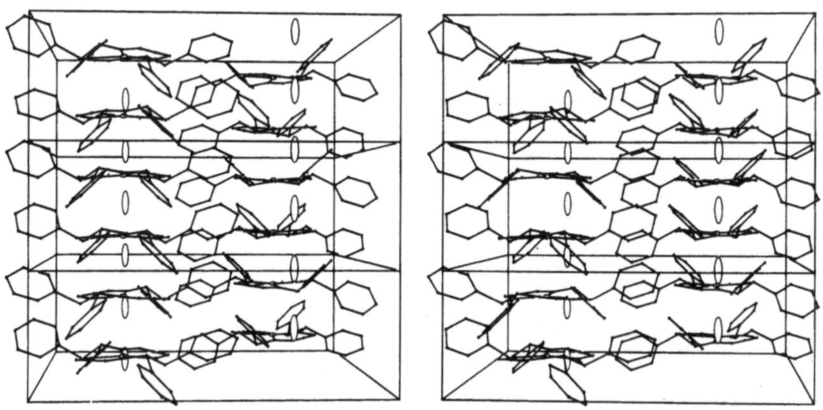

Fig. 23. A stereoview of the unit cell of $Ni(dpg)_2I$. The *a* axis is horizontal to the right, the *b* axis is away from the reader, and the *c* axis is vertical from bottom to top. The vibrational ellipsoids of the Ni and I atoms are drawn at the 50% level. All other atoms are drawn arbitrarily small (Ref. 4)

but by no means inaccessible state of Ni. However, several lines of evidence, then and later, did not support the presence of Ni^{III} in these systems. An X-ray photoelectron spectroscopic study of $Ni(dpg)_2I$[4)] showed no evidence of a trapped valence or of any appreciable change in the charge on the metal upon oxidation. Although early EPR studies suggested a ligand oxidation to yield a free radical[167)], later studies on purer materials found no EPR signal at all[191)]. Foust and Soderberg[167)], on the basis of a limited crystallographic study, deduced that the $M(dpg)_2X$ systems (M = Ni, Pd; X = I, Br) adopted structural motif 1 (Fig. 2a), with M-M distances of 3.26 to 3.36 Å, depending upon M and X, and with channels in which the halogen atoms (nature unknown) were located. Such an arrangement was consistent with the known nature of the starch-iodine complex[192)]. A complete three-dimensional structure determination of $Ni(dpg)_2I$, in combination with spectroscopic studies, has established that this compound contains mainly,

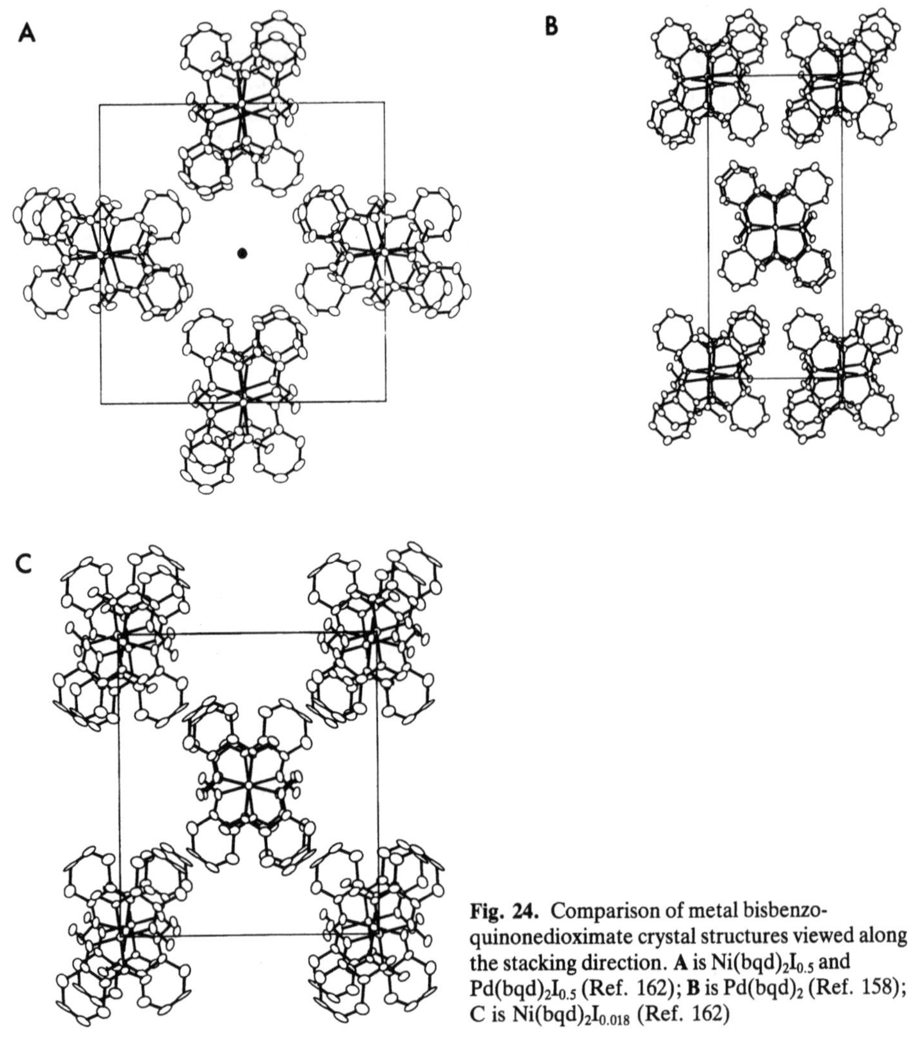

Fig. 24. Comparison of metal bisbenzoquinonedioximate crystal structures viewed along the stacking direction. **A** is $Ni(bqd)_2I_{0.5}$ and $Pd(bqd)_2I_{0.5}$ (Ref. 162); **B** is $Pd(bqd)_2$ (Ref. 158); C is $Ni(bqd)_2I_{0.018}$ (Ref. 162)

if not exclusively, I_5^- ions in the channels and hence that the Ni is in the formal oxidation state of 2.20 on the assumption that the ligand is not oxidized[4]. In addition, this study established that the electrical conductivity in the Ni-Ni direction is about 10^5 times that in the parent complex. Thus partial oxidation converts an insulator to a semiconductor possessing highly anisotropic conductivity properties. The site of partial oxidation in these halogenated compounds remains a mystery, owing in part to the paucity of magnetic data. Miller has suggested an indirect method of determining whether the oxidation is metal or ligand centered based on changes in the relative orientation of adjacent molecules upon halogenation[193]. Oxidation out of a metal orbital, because of its a_g symmetry, should not affect the relative orientation, whereas oxidation out of a ligand orbital should result in a rotation to allow for better ligand-ligand overlap between adjacent molecules and thus a stabilization of the partially occupied molecular orbital.

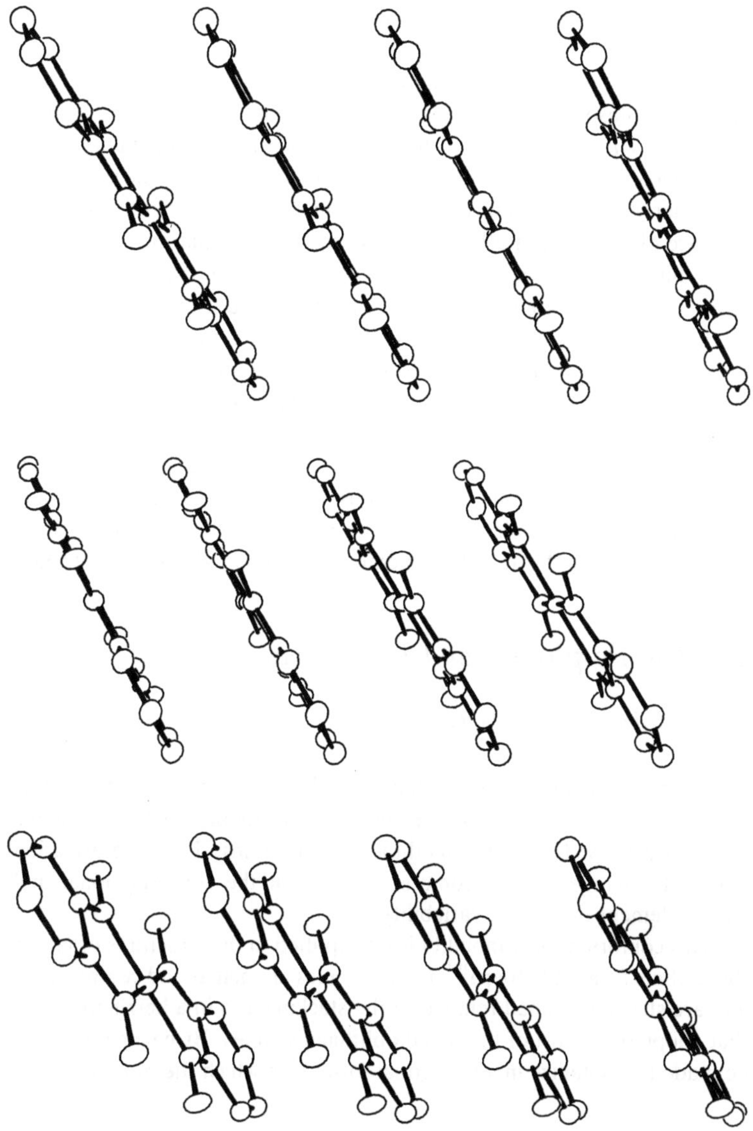

Fig. 25. Packing diagram of $Ni(bqd)_2$ plotted from the data of Ref. 156. Vibrational ellipsoids are drawn at the 50% level (Ref. 162)

On the basis of this argument and the known crystal structures of parents and oxidized products, $Ni(dpg)_2I$ and $Pd(dpg)_2I$ are metal oxidized while $Ni(bqd)_2I_{0.52}$ is ligand oxidized. Clearly, this argument ignores crystal packing considerations and may well be incorrect.

A simple rationalization of the low conductivity displayed by these materials would be that in every case oxidation is metal centered, and that the M-M distances are too great to support metal-like conductivity[6–10]. However, this would require a reinterpretation of the photoelectron spectroscopic data[4].

Are there other manifestations of partial oxidation? The answer is not clear. One can surely note that the few structures known of the partially oxidized materials show structural motif 1, even though some of the parents show motif 2. (In this sense, which is somewhat illusory, partial oxidation brings about a "reduction" in the M-M distance.) But there is no direct evidence that partial oxidation reduces the M-M distance in strictly comparable cases, where parent and oxidized product have the same motif. In particular, note that the Pt-Pt distance of 3.173 Å in $Pt(bqd)_2$ is slightly shorter than the Pd-Pd distance of 3.184 Å in $Pd(bqd)_2I_{0.50}$, despite the fact that the Pt radius is larger than the Pd radius. Included in Tables 8 and 9 are some metrical details on these molecules. Although the data are generally of disappointing accuracy, the effects of oxidation state of the metal (ranging in some instances from +2 to +4) are not evident. The only discernible trend is increasing M-N distances as one descends from Ni to Pd to Pt. Referring again to the conductivity data, we find that the nature of the metal apparently is unimportant. On the other hand, when we examine the porphyrin and phthalocyanine systems we will find upon oxidation by iodine a dramatic increase in conductivity – conductivity that in some instances is metal-like. Clearly the ligand can have an extremely important effect on the transport properties, although such effects are not obvious within the bis(α,β-dionedioximate) compounds of the Ni triad.

4.3 *The Porphyrins and Derivatives*

4.3.1 *Molecular Architecture*

The porphyrinic complexes are all essentially planar. However, the simple porphyrin and tbp macrocycles are somewhat flexible, and can adopt a ruffled, or saddle-shaped distortion through twisting at the methine carbon atoms. These deformations cause almost negligible changes in electronic structure, but the conformational mobility offers an added element of subunit variability.

This conformational mobility is exemplified in the structures of Ni(oep) (Fig. 26). In the triclinic form of Ni(oep)[194] the molecules are planar and centrosymmetric (Fig. 26a) and are stacked inclined relative to the stacking axis in a herringbone pattern similar to that adopted by most of the metallophthalocyanines. By change of solvent, Ni(oep) can be made to crystallize in a tetragonal form[195] in which the molecule has crystallographi-

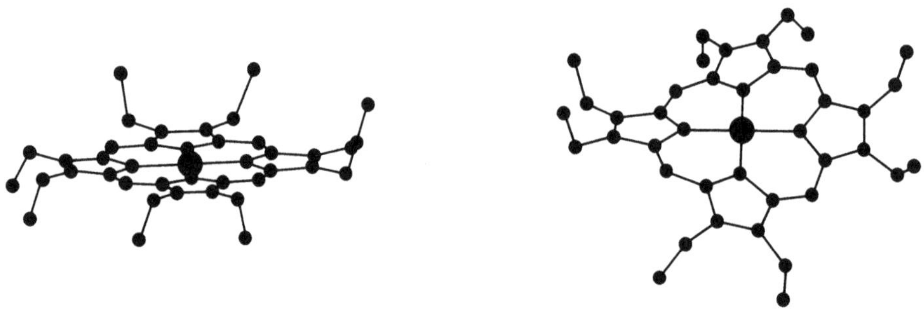

Fig. 26. The Ni(oep) molecule in (*left*, a) triclinic Ni(oep) where the porphyrin ring is planar and (*right*, b) tetragonal Ni(oep) where the porphyrin ring is puckered (Ref. 194)

cally imposed $\bar{4}$ symmetry, and the packing is entirely different from that found for the triclinic form. The pyrrole rings in the tetragonal form of Ni(oep) retain their planarity, but the molecule as a whole is nonplanar because the pyrrole rings are tilted at an angle of 16.4° with respect to the mean molecular plane (Fig. 26b). The interplanar angle between adjacent pyrrole rings is calculated[2] to be ~23° in this ruffled form of Ni(oep) compared with only 2.1° in planar Ni(oep). As a result of the puckering of the Ni(oep) molecule, the β carbon atoms of the pyrrole rings lie as much as 0.5 Å above the mean porphyrin plane. The existence of two conformations of Ni(oep) in crystals without molecules of solvation implies that the energy to deform the porphyrin ring from a planar to a ruffled conformation is not large. A similar mobility is exhibited by the tbp system, as evidenced by the crystal structures obtained for iodine oxidized compounds (*vide infra*).

In contrast, the aza bridges of M(pc) compounds provide rigid connections and the molecules tend more toward planarity. Most four-coordinate metallophthalocyanines have α- and β-polymorphic crystalline forms with the β-polymorph being the more stable and more completely characterized[141)]. This packing motif is very different from the metal-over-metal stacks (Fig. 2a) adopted by the partially oxidized Ni(pc)I_x complex (Fig. 20). Both forms consist of M(pc) slipped-stacks (Fig. 2b) in which the M(pc) molecule is planar and has a crystallographically imposed center of symmetry. The intramolecular distances and angles of the metallophthalocyanine molecules in the α- and β-crystalline modifications are very similar, the major differences being in the crystal packing and in particular in the inclination of the M(pc) molecule relative to the stacking axis. The β-polymorphs of metal-free "H_2", Fe^{II}, Ni^{II}, Mn^{II}, Co^{II}, Cu^{II}, and Zn^{II} phthalocyanines are very similar structurally[3], and a good description of the minor differences in the crystal packing in these systems has been reported by Mason et al.[197)]. The M(pc) molecules in all of these systems stack in a characteristic herringbone pattern as shown in Fig. 27. The angle between the perpendicular to the MN_4 plane and the stacking (b) axis ranges from 45.7° in H_2(pc)[198)] to 48.4° in Zn(pc)[199)]. As a result of the inclination of the M(pc) molecule relative to the stacking axis, the central metal atom of one M(pc) molecule is close enough to an azamethine nitrogen atom in the M(pc) molecules above and below it so that two weak intermolecular M···N bonds are formed, giving the central metal atom a distorted octahedral geometry. The intermolecular M···N bond lengths range from 3.15 Å in Mn(pc)[197)] to 3.28 Å in Cu(pc)[200)]. The isoindole moieties of the pc ring retain their planarity although the azamethine nitrogen atom involved in the intermolecular M···N bond deviates slightly from the mean molecular plane in the direction of the metal atom. The stabilization of the β-polymorph relative to the α-polymorph is attributed to the formation of this intermolecular M···N bond. A metal such as Pt^{II} is less likely to be stabilized by this octahedral coordination, and indeed the β-polymorph of Pt(pc) has not yet been isolated. Instead, α- and γ-polymorphs of

---

2 Meyer[195)] reported the crystal structure of tetragonal Ni(oep) and defines the angle between adjacent pyrrole rings as twice the angle between the plane of one pyrrole ring and the mean molecular plane. We prefer to define this angle as the dihedral angle between the planes of two adjacent pyrrole rings and have calculated this angle using the atomic coordinates given in Ref. 195.

3 The Ni(pc) molecule is not included in any comparisons since the structure of Ni(pc), completed in 1937[196)], is not as accurate as the structures of the other M(pc) systems.

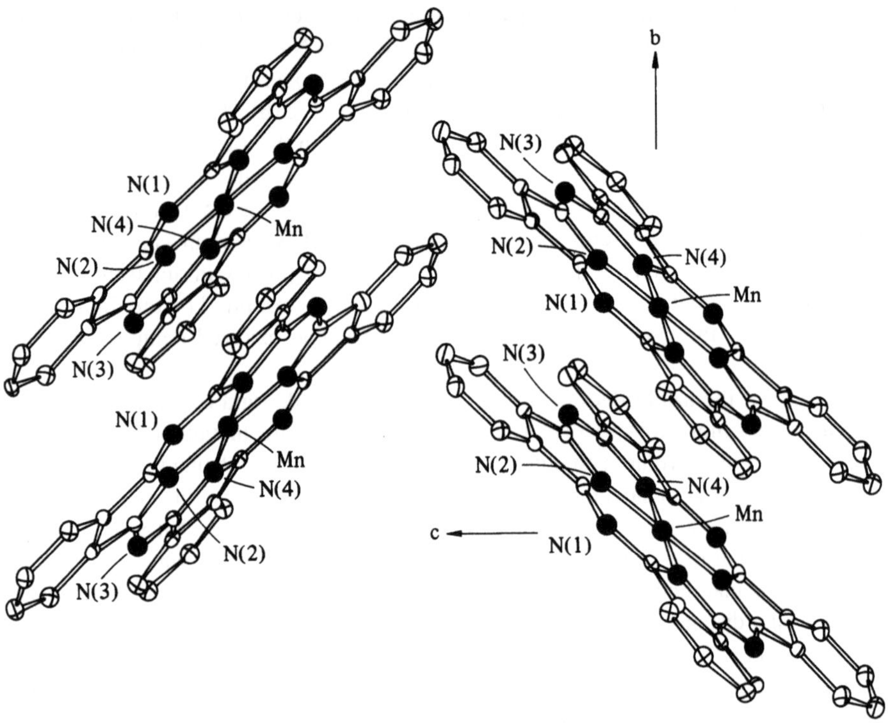

Fig. 27. A view of the herringbone stacking pattern found in β-Mn(pc), a typical β-phthalocyanine (Ref. 197)

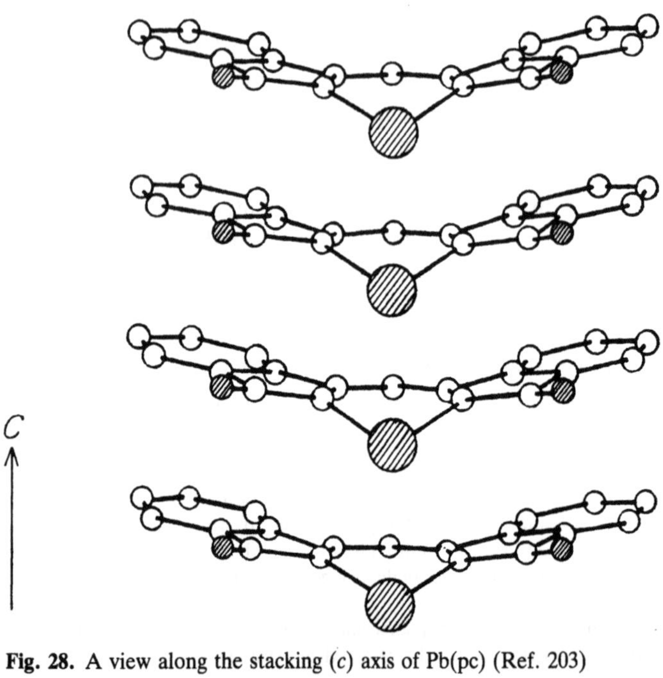

Fig. 28. A view along the stacking (c) axis of Pb(pc) (Ref. 203)

Pt(pc) are stable, and both have been structurally characterized[201, 202]. In these systems, the angle between the perpendicular to the PtN_4 plane and the stacking axis is 25.3 and 30.8° for the α- and γ-polymorphs, respectively, as compared with the ~47° angle found in the β-polymorphs described above. As a result, the intermolecular distance between the Pt atom and the closest azamethine nitrogen atom in the α- and γ-polymorphs is increased (relative to the β-polymorphs) to 3.8 Å.

Of the metallophthalocyanines that crystallize in forms other than the α-, β-, and γ-polymorphs, the most notable are Pb(pc)[203] and Ga(pc)F[130]. The Pb(pc) molecules are stacked metal-over-metal, as shown in Fig. 28, and the Pb-Pb spacing is 3.73 Å. The phthalocyanine ring deviates markedly from planarity, although the separate isoindole moieties retain their planarity. In Ga(pc)F the Ga atoms are symmetrically bridged by F, with a Ga-F distance of 3.92 Å; the pc rings are eclipsed, rather than staggered[130].

4.3.2 Large-Ring Metallomacrocycles

Studies of the compounds prepared by iodination of large-ring metallomacrocycles graphically demonstrate the flexibility in chemical and electronic structure[204] inherent in the use of such complexes to build molecular conductors. The first indication that M(pc) compounds could form the basis for a wide class of molecular conductors was the isolation of conducting polycrystalline $M(pc)I_x$ powders[122]. This result prompted a substantial interest in preparing single crystals of I_2 oxidized, large-ring metallomacrocycles. The first published report was the isolation of single crystals of a pair of materials obtained by the partial oxidation of Ni(omtbp) with iodine[124] and this was followed shortly by a report of the preparation of Ni(pc)I single crystals[3]. Both early reports have been followed by detailed single-crystal studies of structural, magnetic, and charge-transport properties of these compounds[2, 125] and the characterization of numerous related systems. Table 10 lists the key features of the currently known materials. Individual compounds will be described in a fashion most convenient for comparisons.

Table 10. Structure and properties of highly conducting segregated stack, non-integral oxidation state crystals

Compound	Structural data[a]			Properties	Ref.
$Ni(pc)(I_3)_{0.33}$	P 4/mcc Z = 2 R = 0.042 symmetry 4/m	d(Ni-Ni) = d(Ni-N) =	3.244(2) 1.887(6)	PS σ = 260–750 (Type B) $\sigma^{-1}\ \alpha\ T^{1.9}$ T_{MI} = 55 °K	2, 3, 127
$Ni(tbp)(I_3)_{0.33}$	P 4/mcc Z = 2 R = 0.055 symmetry 4/m	d(Ni-Ni) = d(Ni-N) =	3.217(5) (116 °K) 3.26(2) (298 °K) 1.966(9)	PS σ = 150–340 (Type A) T_m = 95 °K $\sigma\ \alpha\ T^{-1}$ (T > 95) Δ = 110 K (T < 95)	128, 132

Table 10 (continued)

Compound	Structural data[a]			Properties	Ref.
$Ni(tatbp)(I_3)_{0.33}$				PS σ = 110–200 (Type A) T_m = 150 °K $\sigma \alpha T^{-1}$ (T > 150) Δ = 0.095 eV	133
$Ni(omtbp)(I_3)_{0.36}$	$P4_2/nbc$ Z = 4 R = 0.050 symmetry $\bar{4}$	d(Ni-Ni) = d(Ni-N) =	3.778(5) 1.953(5)	CL to 10 °K 4–10 °K MT σ = 4–16 (Type A) T_m = 300 °K	124, 125, 126
$Ni(omtbp)(I_3)_{0.97}$				HAF J < 10 cm^{-1} σ = 1–4 (Type A) T_m = 340 °K	124, 125, 126
$Ni(tmp)(I_3)_{0.33}$	P4/nnc Z = 4 R = 0.047 symmetry $\bar{4}$	d(Ni-Ni) = d(Ni-N) =	3.466(3) (114 °K) 1.938(3)	MT at T = 28 °K σ = 40–280 (Type A) T_m = 115 °K	135
$Li_{0.7}[Pt(mnt)_2]$ · 2 H_2O		SS d(Pt-Pt) =	 3.62	σ = 100 (Type A)	120, 121

[a] Unless otherwise noted, the stacking is MOM

4.3.2.1 Ni(omtbp)[124–126]

The two $Ni(omtbp)I_x$ materials isolated following oxidation with iodine have very different stoichiometries, x = 1.08 ± 0.01 and 2.9 ± 0.3, and represent the only case to date in which a porphyrinic metallomacrocycle has yielded crystals at different levels of partial oxidation. The crystal structure for the x = 1.08[125] material exhibits columns of macrocycles, as illustrated in Fig. 2a and 20. The iodine occurs as I_3^- chains which parallel the macrocyclic columns, and which exhibit an interesting one-dimensional disorder. The iodine disorder in $Ni(omtbp)I_{1.08}$ is much more severe than that found in the Ni(pc)I[2] or Ni(tbp)I[132] systems (see below). In order to solve the structure of $Ni(omtbp)I_{1.08}$, the disordered iodine atom had to be treated as a statistical distribution of electron density, with the fit for both cosine and triangle-shaped distributions being comparable (Fig. 29).

The compound $Ni(omtbp)I_{1.08}$ thus exhibits the general topology for one-dimensional conductivity. However, in this compound the macrocycle is "puckered" by intramolecular steric hindrance among the exocyclic methyl groups; it is on a site of $\bar{4}$ symmetry and is severely ruffled (Fig. 30). The isoindole moieties in the Ni(omtbp) molecule are essentially planar but are titled at an angle of ~20° with respect to the mean molecular plane. This large deviation from planarity allows the molecule to avoid close contacts between the bulky methyl substituents on the fused benzene ring of the isoindole unit. As a result of this ruffling of the porphyrin ring, the Ni(omtbp) molecules cannot approach each

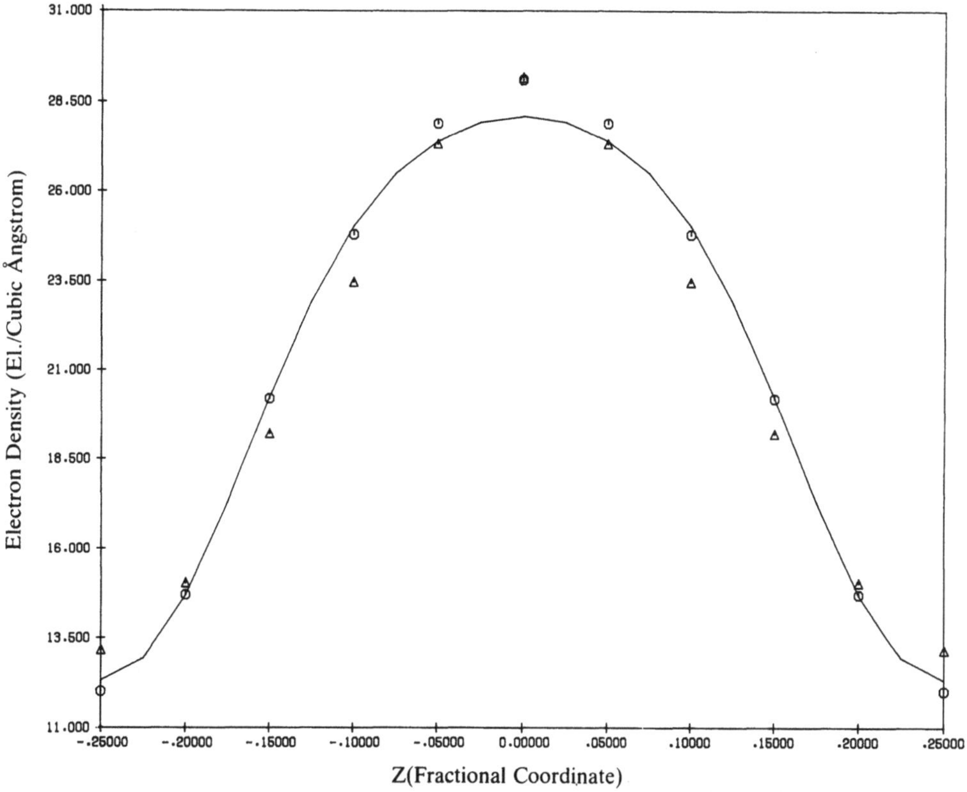

Fig. 29. A plot of observed (—) vs. calculated electron density for Ni(omtbp)$I_{1.08}$ along the line 1/4, 1/4, z: the calculated points are for triangular (△) and cosine (○) distributions of electron density

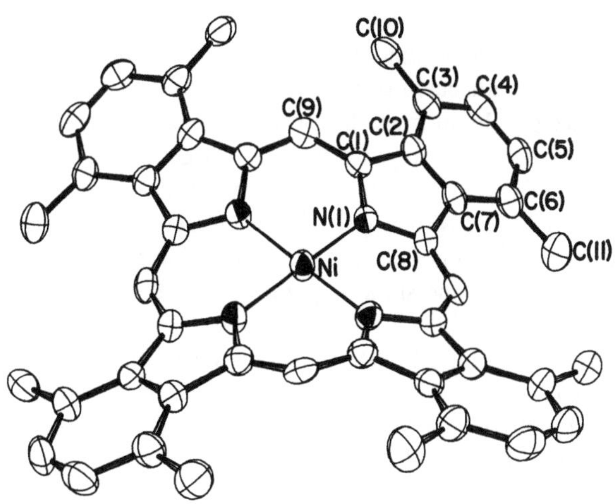

Fig. 30. A perspective drawing of the Ni(omtbp) molecule. Hydrogen atoms have been omitted. (Ref. 125)

other closely. The intrastack spacing, as measured by the distance between adjacent Ni atoms, is 3.778(5) Å[125] at 298 °K, as compared with ca. 3.25 Å in other macrocyclic compounds in non-integral oxidation states. Similarly, the shortest interatomic contacts between adjacent Ni(omtbp) units are also long, 3.46 Å compared with typical values of ca. 3.25 Å[2, 132].

These materials probably offer the most telling example of the critical role that the non-integral oxidation state plays in creating a conductive molecular crystal. The poor intermolecular overlap suggests that the electron transfer matrix element is much smaller than in other conductive molecular crystals. Nevertheless, both Ni(omtbp)I_x materials exhibit moderately high, metal-like conductivities at or above room temperature. These compounds also play an important role in correlating experimental and theoretical studies of charge-transport in conducting molecular crystals[126].

4.3.2.2 Ni(L), L = pc, tbp, tatbp

Iodine oxidation of Ni(pc)[2, 3, 122], Ni(tbp)[128, 132], and Ni(tatbp)[128, 133, 205] in each case leads to isolation of crystals with composition Ni(L)I. Full X-ray diffraction, resonance Raman, and ^{129}I Mossbauer studies show that the first two iodinated materials are isostructural and isoionic, consisting of columnar stacks of planar metallomacrocycles with parallel chains of triiodide counterions (Fig. 20). Furthermore, X-ray photographs of the Ni(tatbp)I[133] system are identical to those of Ni(pc)I and Ni(tbp)I, so all three of these systems appear to be structurally very similar. The electronic properties of the three Ni(L) parent molecules differ significantly because of the different electronegativities of the (–N=) and (–CH=) bridges[204]. Since the Ni(L)I crystals exhibit no concomitant changes in ionicity or structure, they offer an opportunity to examine the dependence of solid state properties on purely electronic characteristics of the subunit in the absence of competing effects. The metallomacrocycle in each of these compounds is on a site of 4/m symmetry and consequently is crystallographically constrained to be planar. Thus, there is no impediment to strong intrastack interactions, and the metal-metal spacing along a column of Ni(pc)I is 3.244(2) Å at 298 °K[2], and the spacings for Ni(tbp)I are 3.26(2) at 298 °K and 3.217(5) Å at 116 °K[132].

Structural parameters for Ni(pc)I, Ni(tbp)I, and Ni(omtbp)$I_{1.08}$ are included in Table 10. Even the major structural differences between the large-ring systems are small, and can be attributed to the expected differences between the phthalocyanine and tetrabenzoporphyrin rings themselves. The larger central hole that is present in the tetrabenzoporphyrin ring is reflected in larger Ni-N bond distances for the Ni(tbp)I and Ni(omtbp)$I_{1.08}$ systems relative to the Ni(pc)I system. Also, the C_a-C_m-C_a angles in Ni(tbp)I[132] and Ni(omtbp)$I_{1.08}$[125] are larger than the corresponding C_a-N_m-C_a angle in Ni(pc)I[2]. This contraction of the azamethine bridge angle in the Ni(pc) system is probably caused by the steric requirements of the lone-pair electrons on the azamethine nitrogen atom[206].

In Ni(pc)I and Ni(tbp)I, the iodine superlattice spacing, c′, is a multiple of the Bragg spacing, c (c′ = 3/2 c), so the superlattice is commensurate with the Bragg lattice and the repeat unit contains three Bragg sites. The disorder in these two systems is not severe, and could be modeled satisfactorily[2, 132, 176]. No special treatment of the iodine atoms was necessary to solve the Bragg structures, and the only structural manifestation of the disorder, aside from diffuse X-ray scattering, is an increase in the thermal parameter of I along the chain direction.

The properties of Ni(tbp)I, Ni(pc)I, and Ni(tatbp)I are characteristic of "molecular metals", but the details of the charge transport properties differ significantly (Fig. 31)[128, 132, 205]. Room-temperature conductivities and, of more significance, mean-free-paths for carrier motion along the stack are comparable with values reported for other one-dimensional "metals", including the organic charge-transfer salts[5, 207–209] (Table 11).

Probably the most unusual result that emerges from this comparison concerns the nature of the charge carriers. In Ni(pc)I, the metal retains its formal Ni^{II} oxidation state, playing a secondary role in the conduction process, and the charge carriers are associated with delocalized π-orbitals on the macrocycle. However, the carrier spin g-values and linewidths (Γ) for Ni(tbp)I and Ni(tatbp)I are anomalous[128, 132, 205] and the results have been interpreted to indicate that Ni(tbp)I and Ni(tatbp)I are the first partially oxidized complexes in which the charge carriers exhibit both metal and ligand properties. The

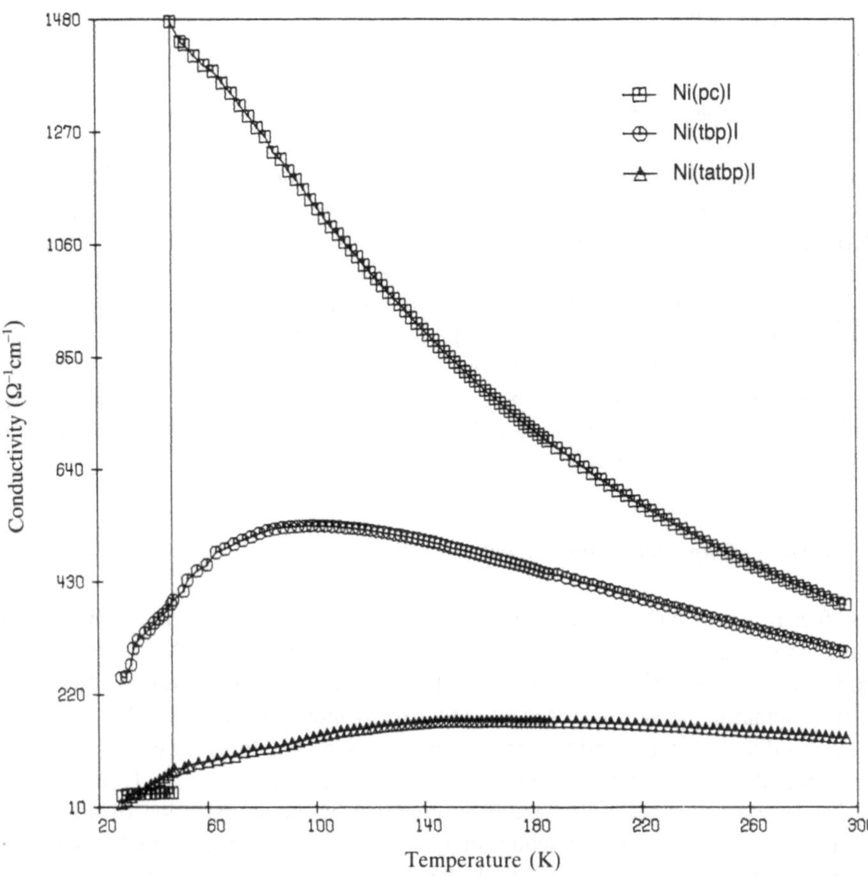

Fig. 31. Temperature-dependent single crystal conductivity of Ni(pc)I, Ni(tbp)I, and Ni(tatbp)I (Refs. 2, 132, 133)

Table 11. Comparison of mean free paths for some organic and inorganic molecular conductors

Compound	RT Conductivity (Ω^{-1} cm^{-1})	Mean free path (intermolecular spacings)	Ref.
TTT_2I_3	1000–10000	2.0–20	207
HMTSF-TCNQ	1400– 2200	1.6–2.5	208
TTF-TCNQ	500– 1000	0.4–0.8	209
KCP	100	0.6	5
Ni(pc)I	260– 750	1.0–2.3	2
Ni(tbp)I	180– 330	0.7–1.3	132
Ni(tatbp)I	110– 200	0.4–0.8	133
Ni(omtbp)$I_{1.08}$	16	0.04	125
Ni(tmp)I	40– 280	0.07–0.51	135

electron "hole" created by iodine oxidation is associated with both π and d "bands", and can "jump" between the metal and the macrocycle, as well as between one unit and its neighbors. In short, these complexes display a novel, doubly mixed valence state, the $[Ni(L)]^+$ cation itself exhibiting interconversion between the electronic tautomers, $[Ni^{II}(L)^+]$ and $[Ni^{III}(L)]$[132].

4.3.2.3 Incompletely Characterized Systems

Single crystals of a number of other iodinated phthalocyanines, tetrabenzoporphyrins, and triazatetrabenzoporphyrins have also been prepared[127, 131, 133], as have crystals of Co(pc)Br[127, 131]. Typically, these compounds have the stoichiometry M(L)$I_{1.0}$, and it is likely that the structures are the same as for Ni(pc)I and Ni(tbp)I. All these compounds conduct well, but with a variety of temperature responses.

The compounds Al(pc)F and Ga(Pc)F form conductive powders, $(M^{III}(pc)F)Y_x$, $Y = I_3^-, BF_4^-, PF_6^-$, upon partial oxidation[129, 130], as do phthalocyanine polymers[136, 137] of the form $(\text{-M(L)-B})_n$, where L = pc or substituted pc, M = Si, Ge, or Sn, and the bridging group B may be O^{-2}, $-O-CH_2-CH_2-O-^{-2}$, or $O-C_6H_4-O^{-2}$ [210, 211]. Related polymers are being prepared[138].

4.3.3 Small-Ring Metallomacrocycles

The first report of high conductivity with small-ring metallomacrocycle is found in the study of the iodine oxidation products of metallooctaethylporphyrins, M(oep)[134]. Iodine oxidation of Ni(oep), Cu(oep), and H_2(oep) yields polycrystalline materials with a range of stoichiometries. Resonance Raman spectroscopy indicates that the iodine occurs as the polyiodide I_5^- species, and thus the materials present an interesting contrast with the large-ring M(L)I_x compounds, where, to date, iodine occurs only as the I_3^- anion.

Single crystals of Ni(tmp)I have been prepared by iodine oxidation of the parent complex[128, 135]. An X-ray diffraction study shows metal-over-metal stacks, with surrounding iodine chains, as is seen with the large-ring systems (Figs. 2a, 20). The flexibility of the porphyrin ring and the consequent futility in trying to anticipate structural details is evident in the Ni(tmp)I system. Despite the fact that both the unoxidized Ni(tmp) molecule[143] and the Ni(tmp) moiety in the charge-transfer complex [Ni(tmp)][TCNQ][44] are planar, the Ni(tmp) molecule in Ni(tmp)I has crystallographically imposed $\bar{4}$ symmetry and is definitely puckered. It is tempting to try to attribute this change in conformation to the oxidation of the porphyrin ring. However, since Ni(oep) exists in both a planar and a ruffled form depending on the solvent used for crystallization[194, 195], and the Ni(pc)[2] and Ni(tbp)[132] systems both retain their planarity upon iodine oxidation, it is most reasonable to attribute the nonplanarity of the porphyrin in Ni(tmp)I to crystal packing forces. It is for such reasons that we have declined to speculate on the crystal packing of the $M(oep)I_x$ compounds.

Structural data for the Ni(tmp)I system are included in Table 10, and similar data for the Ni(oep) and two other Ni(tmp) systems are included in Table 12. As was true for the ruffled Ni(oep) system, the pyrrole rings in Ni(tmp)I retain their planarity but are tilted with respect to the mean molecular plane. The ruffling of the porphyrin results in a decrease in the Ni-N bond distance in Ni(tmp)I relative to that found for Ni(tmp)[143] and [Ni(tmp)][TCNQ][44]. The ruffling of the porphyrin ring has an effect on the crystal packing that is similar to that observed for $Ni(omtbp)I_{1.08}$, where adjacent molecules in the stack cannot approach each other as closely as they might if planarity were maintained. In Ni(tmp)I, the Ni-Ni separation of 3.466(3) Å is 0.3 Å shorter than that found in $Ni(omtbp)I_{1.08}$. The iodine superlattice in Ni(tmp)I is incommensurate with the Bragg lattice ($c' = 1.39c$), and, as was the case with $Ni(omtbp)I_{1.08}$, the disorder is so severe that a statistical distribution of electron density was needed to model it[135].

The conductivity of Ni(tmp)I at room temperature is high ($\sigma \sim 100\ \Omega^{-1}cm^{-1}$) and metal-like, increasing until it reaches a rounded maximum at T ~ 115 °K, then decreasing in an activated manner. The spin susceptibility is temperature independent down to a transition temperature of 28 °K, well below the conductivity maximum. At temperatures below the transition, the susceptibility decreases in an activated fashion, with $\Delta/k \approx 60$ °K. Similar observations with organic conductors $(TMTTF)_2Y$, Y = ClO_4^-, BF_4^-, have been attributed to a Peierls transition[212], but in the present instance the mechanism has yet to be elucidated.

Table 12. Structural data for selected unoxidized small-ring nickel porphyrin systems

Compound	Space group	Z	R	Molecular symmetry	Ni-Ni (Å)	Ni-N (Å)	Ref.
triclinic Ni(oep)	$P\bar{1}$	1	0.045	$\bar{1}$	7.617(2)	1.958(5)	194
tetragonal Ni(oep)	$I4_1/a$	4	0.077 (on F^2)	$\bar{4}$	8.228(6)	1.929(3)	195
Ni(tmp)	$P2_1/c$	2	0.15 (on F^2)	$\bar{1}$	5.648(6)[a]	1.959(10)	143
[Ni(tmp)][TCNQ]	$P2_1/n$	4	0.053	1	–	1.950(4)[b]	44

[a] 123 °K; [b] 116 °K

4.4 Other Metallomacrocycles

Of the many $M(dithiolate)_2^{-q}$ crystals discussed in Sect. 2 and 3, none contained a partially oxidized complex or showed a high conductivity. However, the partially oxidized compound $Li_{1-x}Pt(mnt)_2 \cdot 2\,H_2O$, where $x = 0.25$, has been prepared recently and found to exhibit a metal-like conductivity with a room temperature value of about $200\ \Omega^{-1}cm^{-1}$ [120]. The preliminary observation of high conductivity in the Ni analogue led to the suggestion that the conductivity is at least in part ligand-based, and this is supported by the observation of a large metal-metal distance, 3.65 Å, in the Pt compound[121]. The absence of an observable EPR signal is reminiscent of the integrally oxidized, stacked $M(mnt)_2^-$ systems (Sect. 3.3).

The successful preparation of porphyrinic molecular conductors has recently led to initial studies of analogous metallomacrocycles with the $M(N_4)$, metal-ligand core, the dihydrodibenzotetraaza[14]annulenes and related complexes[172–175].

5 Future Directions

In this review we have discussed the structural properties of DA crystals prepared from planar transition-metal complexes. Although we have not discussed in detail the other characteristics of these crystals, it is obvious that the existence of interesting physical properties is the *raison d'etre* for the structural studies.

Because of the high conductivities and other interesting features of the segregated stack, non-integral oxidation state systems, we expect that major efforts in the near future will focus on their preparation and characterization. Far less effort has gone into the study of these new systems than of organic DA crystals, but it has already been shown that the metal-containing systems exhibit most of the features of importance and interest observed in their organic counterparts. Metal complexes also offer an additional degree of freedom, that of fine-tuning the relative energy of metal and ligand orbitals, and thus offer more chemical flexibility than do the organic DA systems. We anticipate a successful correlation of structures with physical properties that will accelerate the synthesis and utilization of this class of materials.

Acknowledgments. This work has been supported under the NSF-MRL program through the Materials Research Center of Northwestern University (Grant DMR 79-23573) and by National Science Foundation Grants DMR 77-26409 to BMH and CHE 80-09671 to JAI. LJP acknowledges receipt of an NSF Graduate Fellowship. We thank Ms. Jan Goranson and Ms. Carol Lewis for their excellent typing of this review.

6 References

1. Interrante, L. V., Bray, J. W., Hart, H. R., Jr., Jacobs, I. S., Kasper, J. S., Piacente, P. A., Bonner, J. C.: Quasi One-Dimensional Conductors II; Lecture Notes in Physics *96;* Springer-Verlag, New York 1979, p. 55
2. Schramm. C. J., Scaringe, R. P., Stojakovic, D. R., Hoffman, B. M., Ibers, J. A., Marks, T. J.: J. Am. Chem. Soc. *102,* 6702 (1980)

3. Schramm, C. J., Stojakovic, D. R., Hoffman, B. M., Marks, T. J.: Science *200,* 47 (1978)
4. Cowie, M., Gleizes, A., Grynkewich, G. W., Kalina, D. W., McClure, M. S., Scaringe, R. P., Teitelbaum, R. C., Ruby, S. L., Ibers, J. A., Kannewurf, C. R., Marks, T. J.: J. Am. Chem. Soc. *101,* 2921 (1979)
5. Miller, J. S., Epstein, A. J.: Prog. Inorg. Chem. *20,* 1 (1976)
6. Miller, J. S. (Ed.): Extended Linear Chain Compounds; Plenum Press, New York 1981
7. Keller, H. J. (Ed.): Low-Dimensional Cooperative Phenomena; Plenum Press, New York 1975
8. Miller, J. S., Epstein, A. J. (Eds.): Synthesis and Properties of Low-Dimensional Materials, Ann. N. Y. Acad. Sci. *313,* 1978
9. Keller, H. J. (Ed.): Chemistry and Physics of One-Dimensional Metals; Plenum Press, New York 1977
10. Hatfield, W. E. (Ed.): Molecular Metals; Plenum Press, New York 1979
11. Devreese, J. T., Evrard, R. P., Van Doren, V. E. (Eds.): Highly Conducting One-Dimensional Solids; Plenum Press, New York 1979
12. Berlinsky, A. J.: Contemp. Phys. *17,* 331 (1976)
13. Barišić, S., Bjeliš, A., Cooper, J. R., Leontić, B. (Eds.): Quasi One-Dimensional Conductors I, Lecture Notes in Physics *95;* Springer-Verlag, New York 1979
14. Barišić, S., Bjeliš, A., Cooper, J. R., Leontic, B. (Eds.): Quasi One-Dimensional Conductors II, Lecture Notes in Physics *96;* Springer-Verlag, New York 1979
15. Herbstein, F. H.: Perspect. Struct. Chem. *4,* 166 (1971)
16. Alcácer, L. (Ed.): The Physics and Chemistry of Low-Dimensional Solids; D. Reidel Publishing Company, Boston 1980
17. Gutman, F., Lyons, L. E.: Organic Semiconductors; John Wiley & Sons, Inc., New York 1967
18. Hoffman, B. M., Martinsen, J., Pace, L. J., Ibers, J. A. in ref. 6
19. Soos, Z. G.: Ann. Rev. Phys. Chem. *25,* 121 (1974)
20. Soos, Z. G., Klein, D. J.: in Molecular Association (Ed. Foster, R.); Academic Press, New York 1979, Vol. 1, p. 1
21. Williams, R. M., Wallwork, S. C.: Acta Crystallogr. *23,* 448 (1967)
22. Torrance, J. B.: in ref. 8, p. 210
23. Torrance, J. B.: Acc. Chem. Res. *12,* 79 (1979)
24. Bailey, A. S., Williams, R. J. P., Wright, J. D.: J. Chem. Soc., 2579 (1965)
25. Koizumi, S., Iida, Y.: Bull. Chem. Soc. Jpn. *46,* 629 (1973)
26. Prout, C. K., Powell, H. M.: J. Chem. Soc., 4882 (1965)
27. Koizumi, S., Iida, Y.: Bull. Chem. Soc. Jpn. *44,* 1436 (1970)
28. Iida, Y.: Bull. Chem. Soc. Jpn. *44,* 2564 (1971)
29. Kamenar, B., Prout, C. K., Wright, J. D.: J. Chem. Soc. A, 661 (1966)
30. Kamenar, B., Prout, C. K., Wright, J. D.: J. Chem. Soc., 4851 (1965)
31. Bailey, A. S., Prout, C. K.: J. Chem. Soc., 4867 (1965)
32. Murray-Rust, P., Wright, J. D.: J. Chem. Soc. A, 247 (1968)
33. Schmitt, R. D., Wing, R. M., Maki, A. H.: J. Am. Chem. Soc. *91,* 4394 (1969)
34. Wheland, R. C., Gillson, J. L.: J. Am. Chem. Soc. *98,* 3916 (1976)
35. Keller, H. J., Leichert, I., Mégnamisi-Bélombé, M., Nöthe, D., Weiss, J.: Z. Anorg. Allg. Chem. *429,* 231 (1977)
36. Keller, H. J., Leichert, I., Mégnamisi-Bélombé, M., Moroni, W., Nöthe, D., Weiss, J.: Mol. Cryst. Liq. Cryst. *32,* 155 (1976)
37. Mayerle, J. J.: Inorg. Chem. *16,* 916 (1977)
38. King, A. W., Swann, D. A., Waters, T. N.: J. Chem. Soc. Dalton Trans., 1819 (1973)
39. Castellano, E. E., Hodder, O. J. R., Prout, C. K., Sadler, P. J.: J. Chem. Soc. A, 2620 (1971)
40. Cassoux, P., Gleizes, A.: Inorg. Chem. *19,* 665 (1980)
41. Matsumoto, N., Nonaka, Y., Kida, S., Kawano, S., Ueda, I.: Inorg. Chim. Acta *37,* 27 (1979)
42. Grigg, R., Trocha-Grimshaw, J., King, T. J.: J. Chem. Soc., Chem. Commun., 571 (1978)
43. Piovesana, O., Bellitto, C., Flamini, A., Zanazzi, P. F.: Inorg. Chem. *18,* 2258 (1979)
44. Pace, L. J., Ulman, A., Ibers, J. A.: Inorg. Chem. *21,* 199 (1982)
45. Jacobs, I. S., Hart, H. R. Jr., Interrante, L. V., Bray, J. W., Kasper, J. S., Watkins, G. D., Prober, D. E., Wolf, W. P., Bonner, J. C.: Physica *86–88 B,* 655 (1977)

46. Bray, J. W., Hart, H. R., Jr., Interrante, L. V., Jacobs, I. S., Kasper, J. S. Watkins, G. D., Wee, S. H., Bonner, J. C.: Phys. Rev. Lett. *35,* 744 (1975)
47. Kasper, J. S., Interrante, L. V.: Acta Crystallogr. *B32,* 2914 (1976)
48. Jacobs, I. S., Bray, J. W., Hart, H. R., Jr., Interrante, L. V., Kasper, J. S., Watkins, G. D., Prober, D. E., Bonner, J. C.: Phys. Rev. B *14,* 3036 (1976)
49. Delker, G.: Ph. D. Thesis, Univ. of Illinois (1979)
50. Bonner, J. C., Wei, T. S., Hart, H. R., Jr., Interrante, L. V., Jacobs, I. S., Kasper, J. S., Watkins, G. D., Blôte, H. W. J.: J. Appl. Phys. *49,* 1321 (1978)
51. Kasper, J. S., Moncton, D. E.: Phys. Rev. B *20,* 2341 (1979)
52. Moncton, D. E., Birgeneau, R. J., Interrante, L. V., Wudl, F.: Phys. Rev. Lett. *39,* 507 (1977)
53. Ehrenfreund, E., Smith, L. S.: Phys. Rev. B *16,* 1870 (1977)
54. Smith, L. S., Ehrenfreund, E., Heeger, A. J., Interrante, L. V., Bray, J. W., Hart, H. R., Jr., Jacobs, I. S.: Solid State Commun. *19,* 377 (1976)
55. Wei, T. S., Heeger, A. J., Salamon, M. B., Delker, G. E.: Solid State Commun. *21,* 595 (1977)
56. Geiger, W. E., Jr., Maki, A. H.: J. Phys. Chem. *75,* 2387 (1971)
57. Singhabhandhu, A., Robinson, P. D., Fang, J. H., Geiger, W. E., Jr.: Inorg. Chem. *14,* 318 (1975)
58. Vu Dong, Endres, H., Keller, H. J., Moroni, W., Nöthe, D.: Acta Crystallogr. *B33,* 2428 (1977)
59. Endres, H., Keller, H. J., Moroni, W., Nöthe, D.: Acta Crystallogr. *B35,* 353 (1979)
60. Rundle, R. E.: J. Am. Chem. Soc. *76,* 3101 (1954)
61. Yamada, S., Tsuchida, R.: Bull. Chem. Soc. Jpn. *30,* 715 (1957)
62. Hara, Y., Shirotani, I., Onodera, A.: Solid State Commun. *19,* 171 (1976)
63. Wing, R. M., Schlupp, R. L.: Inorg. Chem. *9,* 471 (1970)
64. Jacobs, I. S., Interrante, L. V., Hart, H. R., Jr.: A. I. P. Conf. Proc. *24,* 355 (1975)
65. Mulliken, R. S., Person, W. B.: Molecular Complexes, Wiley-Interscience, New York 1969
66. Chappell, J. S., Bloch, A. N., Bryden, W. A., Maxfield, M., Poehler, T. O., Cowan, D. O.: J. Am. Chem. Soc. *103,* 2442 (1981)
67. Reis, A. H., Jr., Preston, L. D., Williams, J. M., Peterson, S. W., Candela, G. A., Swartzendruber, L. J., Miller, J. S.: J. Am. Chem. Soc. *101,* 2756 (1979)
68. Mayoh, B., Prout, C. K.: J. Chem. Soc. Faraday Trans. 2, 1072 (1972)
69. Mazumdar, S., Soos, Z. G.: Synthetic Metals *1,* 77 (1979/80)
70. Soos, Z. G., Mazumdar, S.: Phys. Rev. B *18,* 1991 (1978)
71. McConnell, H. M., Lynden-Bell, R.: J. Chem. Phys. *36,* 2393 (1962)
72. Chestnut, D. B.: J. Chem. Phys. *45,* 4677 (1966)
73. Pincus, P.: Solid State Commun. *9,* 1971 (1971)
74. Beni, G., Pincus, P.: J. Chem. Phys. *57,* 3531 (1972)
75. Beni, G.: J. Chem. Phys. *58,* 3200 (1973)
76. Dubois, J. Y., Carton, J. P.: J. Phys. (Paris) *35,* 371 (1974)
77. Luther, A., Peschel, I.: Phys. Rev. B *12,* 3908 (1975)
78. Pytte, E.: Phys. Rev. B *10,* 4637 (1974)
79. Poh, G. J., Kommandeur, J.: Mol. Phys. *13,* 373 (1967)
80. Yamada, S.: Bull. Chem. Soc. Jpn. *24,* 125 (1951)
81. Eisenberg, R.: Prog. Inorg. Chem. *12,* 295 (1970)
82. Gray, H. B.: Trans. Met. Chem. *1,* 239 (1965)
83. Livingstone, S. E.: Q. Rev. Chem. Soc. *19,* 386 (1965)
84. McCleverty, J. A.: Prog. Inorg. Chem. *10,* 49 (1968)
85. Hove, M. J., Hoffman, B. M., Ibers, J. A.: J. Chem., Phys. *56,* 3490 (1972)
86. Endres, H., Keller, H. J., Moroni, W., Nöthe, D., Vu Dong: Acta Crystallogr. *B34,* 1823 (1978)
87. Kasper, J. S., Interrante, L. V., Secaur, C. A.: J. Am. Chem. Soc. *97,* 890 (1975)
88. Interrante, L. V., Browall, K. W., Hart, H. R., Jr., Jacobs, I. S., Watkins, G. D., Wee, S. H.: J. Am. Chem. Soc. *97,* 889 (1975)
89. Interrante, L. V., Bray, J. W., Hart, H. R., Jr., Kasper, J. S., Piacente, P. A., Watkins, G. D.: J. Am. Chem. Soc. *99,* 3523 (1977)

90. Interrante, L. V., Bray, J. W., Hart, H. R., Jr., Kasper, J. S., Piacente, P. A., Watkins, G. D.: in ref. 8, p. 407
91. Bray, J. W., Hart, H. R., Jr., Interrante, L. V., Jacobs, I. S., Kasper, J. S., Piacente, P. A., Watkins, G. D.: Phys. Rev. B *16,* 1359 (1977)
92. Forrester, J. D., Zalkin, A., Templeton, D. H.: Inorg. Chem. *3,* 1507 (1964)
93. Davison, A., Edelstein, N., Holm, R. H., Maki, A. H.: Inorg. Chem. *2,* 1227 (1963)
94. Maki, A. H., Edelstein, N., Davison, A., Holm, R. H.: J. Am. Chem. Soc. *86,* 4580 (1964)
95. Schmitt, R. D., Maki, A. H.: J. Am. Chem. Soc. *90,* 2288 (1968)
96. Hamilton, W. C., Bernal, I.: Inorg. Chem. *6,* 2003 (1967)
97. Weiher, J. F., Melby, L. R., Benson, R. E.: J. Am. Chem. Soc. *86,* 4329 (1964)
98. Williams, R., Billig, E., Waters, J. H., Gray, H. B.: J. Am. Chem. Soc. *88,* 43 (1966)
99. Kobayashi, A., Yukiyoshi, S.: Bull. Chem. Soc. Jpn. *50,* 2650 (1977)
100. Perez-Albuerne, N. A., Isett, L. C., Haller, R. K.: J. Chem. Soc. Chem. Commun., 417 (1977)
101. Davison, A., Edelstein, N., Holm, R. H., Maki, A. H.: J. Am. Chem. Soc. *85,* 2029 (1963)
102. Fritchie, C. J., Jr.: Acta Crystallogr. *20,* 107 (1966)
103. Baker-Hawkes, M. J., Billig, E., Gray, H. B.: J. Am. Chem. Soc. *88,* 4870 (1966)
104. Baker-Hawkes, M. J., Dori, Z., Eisenberg, R., Gray, H. B.: J. Am. Chem. Soc. *90,* 4253 (1968)
105. Enemark, J. H., Ibers, J. A.: Inorg. Chem. *7,* 2636 (1968)
106. Lauher, J. W., Ibers, J. A.: Inorg. Chem. *14,* 640 (1975)
107. Cassoux, P., Interrante, L., Kasper, J.: private communication
108. Bekaroglu, Ö., El Sharif, M., Endres, H., Keller, H. J.: Acta Crystallogr. *B32,* 2983 (1976)
109. Beurskens, P. T., Blaauw, H. J. A., Cras, J. A., Steggerda, J. J.: Inorg. Chem. *7,* 805 (1968)
110. Beurskens, P. T., Cras, J. A., van der Linden, J. G. M.: Inorg. Chem. *9,* 475 (1970)
111. Alcácer, L., Maki, A. H.: J. Phys. Chem. *78,* 215 (1974)
112. Alcácer, L., Maki, A. H.: J. Phys. Chem. *80,* 1912 (1976)
113. Alcácer, L., Novais, H., Pedroso, F.: In ref. 10, p. 415
114. Plumlee, K. W., Hoffman, B. M., Ibers, J. A., Soos, Z. G.: J. Chem. Phys. *63,* 1926 (1975)
115. Plumlee, K. W.: Ph. D. Thesis, Northwestern Univ. 1975
116. Plumlee, K. W., Hoffman, B. M., Ratajack, M. T., Kannewurf, C. R.: Solid State Commun. *15,* 1651 (1974)
117. Davison, A., Howe, D. V., Shawl, E. T.: Inorg. Chem. *6,* 458 (1967)
118. Hoffman, B. M.: unpublished
119. Senftleber, F. C., Geiger, W. E., Jr.: Inorg. Chem. *17,* 3615 (1978)
120. Underhill, A. E., Ahmad, M. M.: J. Chem. Soc. Chem. Commun., 67 (1981)
121. Underhill, A. E., Ahmad, M. M.: Mol. Cryst. Liq. Cryst., in press
122. Petersen, J. L., Schramm, C. J., Stojakovic, D. R., Hoffman, B. M., Marks, T. J.: J. Am. Chem. Soc. *99,* 286 (1977)
123. Hoffman, B. M., Phillips, T. E., Schramm, C. J., Wright, S. K.: In ref. 10, p. 393
124. Phillips, T. E., Hoffman, B. M.: J. Am. Chem. Soc. *99,* 7734 (1977)
125. Phillips, T. E., Scaringe, R. P., Hoffman, B. M., Ibers, J. A.: J. Am. Chem. Soc. *102,* 3435 (1980)
126. Hoffman, B. M., Phillips, T. E., Soos, Z. G.: Solid State Commun. *33,* 51 (1980)
127. Schramm, C. J.: Ph. D. Thesis, Northwestern Univ. 1979
128. Euler, W. B., Pace, L. J., Martinsen, J., Hoffman, B. M., Ibers, J. A.: Mol. Cryst. Liq. Cryst., in press
129. Nohr, R. S., Kuznesof, P. M., Wynne, K. J., Kenney, M. E., Siebenman, P. G.: J. Am. Chem. Soc. *103,* 4371 (1981)
130. Brant, P., Nohr, R. S., Weber, D. C., Wynne, K. J.: Mol. Cryst. Liq. Cryst., in press
131. Martinsen, J., Schramm, C. J., Euler, W. B., Pace, L. J., Hoffman, B. M., Ibers, J. A.: in preparation
132. Martinsen, J., Pace, L. J., Phillips, T. E., Hoffman, B. M., Ibers, J.: J. Am. Chem. Soc. *104,* 83 (1982)
133. Euler, W. B., Martinsen, J., Pace, L. J., Hoffman, B. M., Ibers, J. A.: in preparation
134. Wright, S. K., Schramm, C. J., Phillips, T. E., Scholler, D. M., Hoffman, B. M.: Synthetic Metals *1,* 43 (1979/80)

135. Pace, L. J., Martinsen, J., Ulman, A., Hoffman, B. M., Ibers, J. A.: in preparation
136. Schoch, K. F., Jr., Kundalkar, B. R., Marks, T. J.: J. Am. Chem. Soc. *101,* 7071 (1979)
137. Hanack, M., Schneider, O., Metz, J., Kobel, W.: Mol. Cryst. Liq. Cryst., in press
138. Hanack, M., Mitulla, K., Pawlowski, G., Subramanian, L. R.: J. Organomet. Chem. *204,* 315 (1981)
139. Smith, K. M. (Ed.): Porphyrins and Metalloporphyrins, Amer. Elsevier Publ. Comp., New York 1975
140. Dolphin, D. (Ed.): The Porphyrins, Academic Press, New York 1978
141. Lever, A. B. P.: Adv. Inorg. Chem. Radiochem. *7,* 27 (1965)
142. Jackson, A. H.: In ref. 140, Vol. I, p. 365
143. Ulman, A., Gallucci, J., Fisher, D., Ibers, J. A.: J. Am. Chem. Soc. *102,* 6852 (1980)
144. Schramm, C. J., Hoffman, B. M.: Inorg. Chem. *19,* 383 (1980)
145. Calleri, M., Ferraris, G., Viterbo, D.: Inorg. Chim. Acta *1,* 297 (1967)
146. Calleri, M., Ferraris, G., Viterbo, D.: Acta Crystallogr. *22,* 468 (1967)
147. Murmann, R. K., Schlemper, E. O.: Acta Crystallogr. *23,* 667 (1967)
148. Ferraris, G., Viterbo, D.: Acta Crystallogr. *B25,* 2066 (1969)
149. Godycki, L. E., Rundle, R. E.: Acta Crystallogr. *6,* 487 (1953)
150. Williams, D. E., Wohlauer, G., Rundle, R. E.: J. Am. Chem. Soc. *81,* 755 (1959)
151. Gomm, P. S., Thomas, T. W., Underhill, A. E.: J. Chem. Soc. A, 2154 (1971)
152. Bowers, R. H., Banks, C. V., Jacobson, R. A.: Acta Crystallogr. *B28,* 2318 (1972)
153. Frasson, E., Panattoni, C.: Acta Crystallogr. *13,* 893 (1960)
154. Frasson, E., Panattoni, C., Zannetti, R.: Acta Crystallogr. *12,* 1027 (1959)
155. Endres, H., Keller, H. J., van de Sand, H., Vu Dong: Z. Naturforsch. *33 b,* 843 (1978)
156. Leichert, I., Weiss, J.: Acta Crystallogr. *B31,* 2877 (1975)
157. Brill, J. W., Mégnamisi-Bélombé, M., Novotny, M.: J. Chem. Phys. *68,* 585 (1978)
158. Leichert, I., Weiss, J.: Acta Crystallogr. *B31,* 2709 (1975)
159. Endres, H., Mégnamisi-Bélombé, M., Little, W. A., Wolfe, C. R.: Acta Crystallogr. *B35,* 169 (1979)
160. Mégnamisi-Bélombé, M.: J. Solid State Chem. *27,* 389 (1979)
161. Mégnamisi-Bélombé, M.: J. Solid State Chem. *22,* 151 (1977)
162. Brown, L. D., Kalina, D. W., McClure, M. S., Schultz, S., Ruby, S. L., Ibers, J. A., Kannewurf, C. R., Marks, T. J.: J. Am. Chem. Soc. *101,* 2937 (1979)
163. Endres, H., Keller, H. J., Moroni, W., Weiss, J.: Acta Crystallogr. *B31,* 2357 (1975)
164. Endres, H., Keller, H. J., Mégnamisi-Bélombé, M., Moroni, W., Pritzkow, H., Weiss, J., Comès, R.: Acta Crystallogr. *A32,* 954 (1976)
165. Endres, H., Mégnamisi-Bélombé, M., Keller, H. J., Weiss: Acta Crystallogr. *B32,* 457 (1976)
166. Pritzkow, H.: Z. Naturforsch. *31 b,* 401 (1976)
167. Foust, A. S., Soderberg, R. H.: J. Am. Chem. Soc. *89,* 5507 (1967)
168. Endres, H., Keller, H. J., Lehmann, R., Weiss, J.: Acta Crystallogr. *B32,* 627 (1976)
169. Kalina, D. W., Lyding, J. W., Ratajack, M. T., Kannewurf, C. R., Marks, T. J.: J. Am. Chem. Soc. *102,* 7854 (1980)
170. Underhill, A. E., Watkins, D. M., Pethig, R.: Inorg. Nucl. Chem. Lett. *9,* 1269 (1973)
171. Edelman, L. E.: J. Am. Chem. Soc. *72,* 5765 (1950)
172. Lin, L.-S., Marks, T. J., Kannewurf, C. R., Lyding, J. W., McClure, M. S., Ratajack, M. T., Whang, T.-C.: J. Chem. Soc., Chem. Commun. 954 (1980)
173. Wuu, Y.-M., Peng, S.-M.: J. Inorg. Nucl. Chem. *42,* 205 (1980)
174. Wuu, Y.-M., Peng, S.-M., Chang, H.: J. Inorg. Nucl. Chem. *42,* 839 (1980)
175. Hatfield, W. E.: private communication
176. Scaringe, R. P., Ibers, J. A.: Acta Crystallogr. *A35,* 803 (1979)
177. Teitelbaum, R. C., Ruby, S. L., Marks, T. J.: J. Am. Chem. Soc. *100,* 3215 (1978)
178. Teitelbaum, R. C., Ruby, S. L., Marks, T. J.: J. Am. Chem. Soc. *102,* 3322 (1980)
179. Mertes, K. B., Ferraro, J. R.: J. Chem. Phys. *70,* 646 (1979)
180. Jircitano, A. J., Timken, M. D., Mertes, K. B., Ferraro, J. P.: J. Am. Chem. Soc.*101,* 7661 (1979)
181. Jircitano, A. J., Colton, M. C., Mertes, K. B.: Inorg. Chem. *20,* 890 (1981)
182. Wolberg, A., Manassen, J.: J. Am. Chem. Soc. *92,* 2982 (1970)
183. Lever, A. B. P., Wilshire, J. P.: Can. J. Chem. *54,* 2514 (1976)

184. Ref. 140, Vol. V, Chapters 1, 3, 4
185. Bobrovskii, A. P., Sidorov, A. N.: J. Struct. Chem. (Engl. Tranl.) *17,* 50 (1976)
186. Tomkiewicz, Y., Taranko, A. R., Torrance, J. B.: Phys. Rev. Lett. *36,* 751 (1976)
187. Mégnamisi-Bélombé, M.: Z. Anorg. Allg. Chem. *473,* 196 (1981)
188. Kubel, F., Strähle, J.: Z. Naturforsch. *36b,* 441 (1981)
189. Endres, H., Keller, H. J., Lehmann, R., Poveda, A., Rupp, H. H., van de Sand, H.: Z. Naturforsch. *32b,* 516 (1977)
190. Egneus, B.: Anal. Chim. Acta *48,* 291 (1969)
191. Mehne, L. F., Wayland, B. B.: Inorg. Chem. *14,* 881 (1975)
192. Reddy, J. W., Knox, K., Robin, M. B.: J. Chem. Phys. *40,* 1082 (1964)
193. Miller, J. S.: Inorg. Chem. *16,* 957 (1977)
194. Cullen, D. L., Meyer, E. F., Jr.: J. Am. Chem. Soc. *96,* 2095 (1974)
195. Meyer, E. F., Jr.: Acta Crystallogr. *B28,* 2162 (1972)
196. Robertson, J. M., Woodward, I.: J. Chem. Soc., 219 (1937)
197. Mason, R., Williams, G. A., Fielding, P. E.: J. Chem. Soc. Dalton Trans. 676 (1979)
198. Robertson, J. M.: J. Chem. Soc., 1195 (1936)
199. Scheidt, W. R., Dow, W.: J. Am. Chem. Soc. *99,* 1101 (1977)
200. Brown, C. J.: J. Chem. Soc. A, 2488 (1968)
201. Robertson, J. M., Woodward, I.: J. Chem. Soc., 36 (1940)
202. Brown, C. J.: J. Chem. Soc. A, 2494 (1968)
203. Ukei, K.: Acta Crystallogr. *B29,* 2290 (1973)
204. Gouterman, M.: In ref. 140, Vol. III, p. 1
205. Euler, W. B., Hoffman, B. M.: in preparation
206. Hoard, J. L.: In ref. 139, p. 321
207. Hilti, B., Mayer, C. W.: Helv. Chim. Acta *61,* 501 (1978)
208. Bloch, A. N., Cowan, D. O., Beckgaard, K., Pyle, R. E., Banks, R. H., Poehler, T. O.: Phys. Rev. Lett. *34,* 1561 (1975)
209. Thomas, G. A., Schafer, D. E., Wudl, F., Horn, P. M., Rimai, D., Cook, J. W., Glocker, D. A., Skove, M. J., Chu, C. W., Groff, R. P., Gillson, J. L., Wheland, R. C., Melby, L. R., Salamon, M. B., Craven, R. A., De Pasquali, G., Bloch, A. N., Cowan, D. O., Walatka, V. V., Pyle, R. E., Gemmer, R., Poehler, T. O., Johnson, G. R., Miles, M. G., Wilson, J. D., Ferraris, J. P., Finnegan, T. F., Warmack, R. J., Raaen, V. F., Jerome, D.: Phys. Rev. B *13,* 5105 (1976)
210. Esposito, J. N., Sutton, L. E., Kenney, M. E.: Inorg. Chem. *6,* 1116 (1967)
211. Meyer, G., Hartmann, M., Wöhrle, D.: Makromol. Chem. *176,* 1919 (1975)
212. Delhaes, P., Coulon, C., Amiell, J., Flandrois, S., Toreilles, E., Fabre, J. M., Giral, L.: Mol. Cryst. Liq. Cryst. *50,* 43 (1979)

The Chemistry of Technetium: Toward Improved Diagnostic Agents

Michael J. Clarke and Paul H. Fackler

Department of Chemistry, Boston College, Chestnut Hill, MA, 02167, USA

Within the last five years significant advances have been made in elucidating the chemistry of technetium. While this element is continually produced in some abundance during the operation of fission nuclear reactors, its practical application has been limited to nuclear medicine. Owing to the development of the "technetium generator" almost two decades ago, ^{99m}Tc is now the preferred isotope for many diagnostic studies involving the imaging of internal organs by radioscintigraphic techniques; however, many of the agents employed have yet to be well characterized from a chemical standpoint. Owing to its position in the middle of the periodic table, the chemistry of this element has already proven rich, even when limited to aqueous solution. Subtle chemical differences often control the relative stabilities of the various oxidation states and each oxidation state exhibits more than one geometry. Coordination complexes involving Tc(III), Tc(IV) and Tc(V) tend to predominate in preparations used for diagnosis. Recent synthetic advances now pave the way for new series of technetium compounds with well-described chemistry. Of these, methods for coordinating chelating nitrogen ligands to Tc(V) offer substantial versatility. (This review covers the literature through August, 1981.)

Structure and Bonding 50

Abbreviations

acac	2,4-pentanedione
nBu	n-butyl
cat	1,2-dihydroxybenzene
cyclam	1,4,8,11-tetraazacyclotetradecane
diars	o-phenylenebis(dimethylarsine)
diphos	bis(diphenylphosphine)ethane
DMF	N,N-dimethylformamide
DMG	dimethylglyoxime
DTPA	diethylenetriaminepentaacetate
EDTA	ethylenediaminetetraacetate
ema	N,N′-ethylenebis (2-mercaptoacetimide)
en	ethylenediamine
Et	ethyl
HEDP	(1-hydroxyethylidene)diphosphate
HIDA	N-(2,6-dimethylphenylcarbamoylmethyl)-iminodiacetate
HTFMS	trifluoromethanesulfonic acid
IDA	iminodiacetate
Im	imidazole
MDP	methylenediphosphonate
Me	methyl
NADH	reduced nicotinamide adenine dinucleotide
NHE	normal hydrogen electrode
Ph	phenyl
Pyr	pyridine
SCE	standard calomel electrode
TBAP	tetrabutylammonium perchlorate
tdt	3,4-dimercaptotoluene
TEAP	tetraethylammonium perchlorate
TMAP	tetramethylammonium perchlorate

A. Introduction

1. Radiochemical Properties

From 1937, when technetium was re-introduced into the terrestrial environment as the first man-made element[1, 2], until the ready availability of ^{99m}Tc for medical purposes[3], the chemistry of this element was regarded largely as an academic curiosity. Since it is formed in good yield in fission reactors fueled by ^{235}U[4], much early work was devoted to its separation from spent nuclear wastes[4, 5].

The instability of all technetium isotopes with respect to radioactive decay has been a significant deterrent to its systematic chemical investigation. Most chemical studies now utilize the relatively long-lived ^{99}Tc, which is available in quantity. Since this isotope emits essentially only a 292 KeV beta ray, milligram quantities can be handled with negligible radiation exposure. Other technetium isotopes include mass numbers 92 through 105 and 107. Isotopes with mass numbers 92–97 undergo decay by electron capture or positron emission, while those with mass numbers 98–107 emit beta particles and usually produce stable ruthenium isotopes. Isotopes available from commercial sources include ^{95m}Tc (60 day half-life), ^{99m}Tc (6 h), and ^{99}Tc (2.12×10^5 years).

2. Use in Nuclear Medicine

The radiophysical properties of the ^{99m}Tc nuclide make it optimally suited for medical gamma ray scanning[6–9]. Relative to other gamma emitting nuclides, ^{99m}Tc possesses the advantages of: 1) a short half-life (6.03 h) so that it has a high specific activity and decays rapidly enough not to present a long-term hazard, 2) emitting no particulate or extremely high energy radiation, which minimizes the absorbed radiation dose to the patient, 3) decay to a daughter isotope (^{99}Tc), which gives a negligible radiation dose, and 4) emission of an easily collimated gamma ray of sufficient energy (140 KeV) to be detected through layers of tissue.

Concentration of ^{99m}Tc in a lesion causes gamma rays to be emitted from the area so as to produce an excellent positive image of the affected tissue. By monitoring the uptake and excretion of a tagged radioactive molecule as a function of time, radioscintigraphy provides a view of the physiological function of the organ as well as its structure. With the advent of the ^{99m}Tc generator in the late 1960's it became possible to secure this radionuclide conveniently in a pure form[3]. The ^{99m}Tc is produced by the decay of ^{99}Mo absorbed on an alumina column. Separation is usually accomplished by aqueous elution of ^{99m}Tc from the column as the pertechnetate ion, $[TcO_4]^-$. Direct use of the pertechnetate ion is advantageous in the imaging of some types of lesions, particularly those of the thyroid[10, 11] and brain[12, 13]. However, it was not until coordination complexes of this nuclide became available that its use proliferated. The use of ^{99m}Tc for diagnosis has now proven to be so successful that presently approximately 10% of all hospitalized patients are evaluated with a derivative containing this isotope[4].

A variety of technetium imaging agents are presently available. The majority of those used clinically are produced through commercial derivatization kits, which usually involve a reductant in the presence of a complexing agent[15, 16]. These radiopharmaceuticals have been categorized as belonging to two broad classes[15, 17]:

1. Those in which the labeled compound retains essentially the same biodistribution as the untagged molecule. Most agents in this category involve cells, colloids or macromolecules so that the presence of the technetium represents only a small perturbation.
2. Agents in which the biodistribution depends on the properties of the complex itself and on the nature of the particular technetium ion present. This class encompasses nearly all of the smaller technetium coordination complexes used for diagnostic purposes.

Agents of the following are (or have been) fairly commonly used to image various organs or to evaluate specific physiological functions:

1) Phosphates, phosphonates, and iminodiphosphonates for skeletal imaging[18-22].
2) Human serum albumin (HSA), for blood pool assessment, lung, and heart imaging[15, 23].
3) Human erythrocytes, for blood pool imaging, red blood cell mass determination, detection of vascular malformations and nuclear cardiology. Heat-damaged red blood cells also yield an excellent agent for spleen imaging[15, 24].
4) Sulfur colloids, for imaging the reticuloendothelial systems of liver, spleen and bone marrow[15].
5) Micro and macroaggregated spheres of HSA, for evaluation of the lung arteriolar capillary bed[15].
6) Antimony sulfide, for lymphoscintigraphy[25].
7) Phosphates, phosphonates, EDTA, DTPA and methyliminodiacetic acid for the evaluation of renal function. Gluconate, glucoheptonate, ascorbate, insulin and dimercaptosuccinate to image kidney structure[26].
8) Pyrophosphate, glucoheptonate, and tetracycline for infarct visualization[27-29].
9) Iminodiacetate derivatives, especially acetanilido-IDA ligands, such as HIDA; also pyridoxylidenaminates and dihydrothioctic acid, as hepatobiliary agents[30, 31].

Many of these agents have only recently begun to be adequately characterized with respect to their structure and chemical reactivity. This may be due to the fact that only nanogram quantities are required for adequate imaging and, since these amounts are insufficient for most chemical measurements and do not represent an obvious chemical hazard to patients, many medical researchers have not pursured the problem extensively.

Most derivatized ^{99m}Tc radiopharmaceuticals contain the metal in the III, IV, or V oxidation states. While the synthesis of these agents most often involves the reduction of pertechnetate with stannous ion[32-34], other reductants have also been shown to be effective[15, 35-43]. Multiple products often result from the reduction of $[TcO_4]^-$, depending on the redox potential of the particular reductant employed, as well as the concentration of available complexing groups and their affinity for the lower technetium oxidation states. Synthetic reactions often tend to form the insoluble polymer, $[TcO_2]_n$, due to the inherent thermodynamic stability of dioxides in this region of the periodic table.

3. *Tumor-Imaging Agents*

Radioscintigraphic agents have proven useful in detecting the presence of neoplasms, locating growths for biopsy, and imaging metastases in the earliest stages of development[6-12, 15, 44-47]. Although presently less specific than endoscopy, biopsy, and some laboratory tests, nuclear medicine diagnostic procedures are, in contrast, highly sensitive, easily performed, and associated with much lower morbidity. Moreover, when dose considerations are adequately taken into account, they can be repeated safely as often as necessary to assess the efficacy of a therapeutic regimen[7, 8]. Bone scanning has become an important use of ^{99m}Tc agents in the management of breast cancer[48–50]. Such scans are a sensitive initial screen for bony metastases and frequently will detect involvement of the skeletal system in 30–50% of patients with normal radiographic studies on the same region[48–50]. Pertechnetate ion and various ^{99m}Tc phosphate complexes have also been used in the detection of a variety of soft-tissue cancers[51, 52].

Various mechanisms for tumor uptake of radiopharmaceuticals have recently been reviewed[6, 44, 53]. In addition it may be possible to take advantage of known aspects of tumor metabolism and coordination chemistry to develop new agents of enhanced specificity. Tumor cells generally grow more rapidly than those of normal tissue with a concommitant greater uptake of glucose and oxygen[54–60]. In fact, tumor tissue usually develops so rapidly that the process of neovascularization cannot keep pace to adequately supply the growing tumor[56]. This results in areas of hypoxia even at only micrometers from blood capillaries[61]. Moreover, since the bulk of the oxygen content is depleted from the blood in tumors at a fraction of the distance along the capillary and capillaries tend to be longer in larger tumors, oxygen content decreases with increasing tumor size and the central areas of larger tumors tend to be anoxic[56].

Due to depressed respiration caused by tissue hypoxia, glycolysis assumes the primary energy producing role in rapidly growing malignancies[63]. This results in increased levels of lactic acid and concommitant lower pH[55]. Tissue hypoxia coupled with the glycolytic production of NADH causes the interior of tumors to be more reducing and more acidic than the surrounding normal tissue[64]. In addition to their intracellular electron transfer proteins, some varieties of tumor cells contain an NADH oxidoreductase on their external surfaces, which is also capable of reducing metal ions[62].

Technetium complexes combining ligands which are all firmly bound in the higher oxidation state, but only some of which are firmly bound in the lower oxidation state should allow the metal to substitute new ligands on reduction, and so cause the metal to bind and become fixed in the reducing environment of tumor tissue. This approach has already met with some success in the design of anticancer drugs containing ruthenium[65–67], whose chemistry is somewhat analogous to that of technetium[68]. For those redox couples which are pH dependent, the lower pH of tumor tissue may increase the reduction potential by as much as 30 mV over the corresponding reduction potential in blood[55–57, 67]. Since decreased pH enhances the hydrolysis of some technetium complexes, the resulting TcO_2 and other sparingly soluble products may be preferentially deposited in neoplasms. It must also be assumed that the increased temperatures[60] present in many types of tumor tissue will cause these reactions to occur slightly more rapidly than in normal tissue.

B. The Chemistry of Technetium

1. Technetium(VII)

The Tc(VII) compounds Tc_2O_7 and Tc_2S_7 are both volatile (0.1 mm Hg at 100 °C for the oxide) and can be lost upon heating[5]. Technetium heptoxide is a yellow crystalline solid consisting of TcO_4 tetrahedra which are bridged through linear Tc-O-Tc links[69]. On dissolution in water it yields the also volatile $HTcO_4$. The Na(I), K(I), Cs(I), Ag(I)[70], and $(NH_4)^+$ [71,72] salts of $[TcO_4]^-$ have been characterized. The stable ammonium salt, which can be dried at temperatures above 100 °C, is most commonly utilized for synthetic and analytical purposes. The structure of this salt has been precisely determined as a function of temperature and reveals the Tc-O bond lengths in this tetrahedral ion to be 1.72 Å[72]. The heptasulfide is insoluble and is widely employed in various processes for the isolation of Tc; however, its precipitation is often not quantitative[4,5,73]. Other Tc(VII) compounds include TcO_3F[74,75] and TcO_3Cl[76,77], which are liquids at room temperature.

2. *Technetium(VI)*

Electrochemical reduction of $[TcO_4]^-$ in acetonitrile solution containing tetramethylammonium salts yields violet crystals of $[(CH_3)_4N]_2(TcO_4)$, which are isotypic with the rhenium and manganese analogs[78]. Electrochemical evidence also exists for the transient formation of $[TcO_4]^{2-}$ from pertechnetate in aqueous alkaline media[79]. Other Tc(VI) compounds reported to date include: TcF_6, $TcCl_6$[4], TcO_3, $TcOF_4$[75], and $TcOCl_4$[77]. The compound $TcOF_4$ is the only example of a trimeric technetium species. In the crystalline form, octahedrally coordinated units are linked into trimers by cis bridging fluorine atoms[80]. The two terminal Tc-F distances are 1.81 Å, while the bridging Tc-F distances are 2.26 Å and 1.89 Å. The Tc-O distance opposite the longer Tc-F bond is 1.66 Å.

Table 1. Reduction potentials for selected technetium compounds

Couple	$E_{1/2}$ (V vs. NHE)[a]	Media	Ref.
$[TcO_4]^{1-,2-}$	−0.61	0.1 M NaOH	79
$[TcO_4]^{1-,2-}$	−1.60	0.05 M TMAP in CH_3CN	170
$[TcO(SCH_2COS)_2]^{1-,2-}$	−1.11	0.1 M TBAP in CH_3CN	91
$[TcO(SOCCOS)_2]^{1-,2-}$	−0.51	0.1 M TBAP in CH_3CN	91
$[TcO(S_2C_2(CN)_2))_2]^{1-,2-}$	−0.40	0.1 M TBAP in CH_3CN	91
$[TcO(tdt)_2]^{1-,2-}$	−1.28	0.1 M TBAP in CH_3CN	91
$[Tc(NCS)_6]^{2-,3-}$	0.42	0.1 M TBAP in CH_3CN	131
$[TcCl_4(PMe_2Ph)_2]^{0,1-}$	−0.20	0.1 M TBAP in CH_3CN	159
$[TcCl_2(PMePh)_3(MeCN)]^{1+,0}$	0.50	0.1 M TBAP in CH_3CN	160
$[TcCl_2(PMe_2Ph)_3(MeCN)_2]^{0,1-}$	−0.57	0.1 M TBAP in CH_3CN	160
$[TcCl(PMe_2Ph)_3(MeCN)_2]^{1+,0}$	0.07	0.1 M TBAP in CH_3CN	159
$[TcCl_2(diphos)_2]^{1+,0}$	0.20	0.5 M TEAP in DMF	128
$[TcCl_2(diars)_2]^{1+,0}$	0.15	0.5 M TEAP in DMF	128
$[TcI_2(diars)_2]^{1+,0}$	0.34	0.5 M TEAP in DMF	128
$[TcCl_2(diars)_2]^{0,1-}$	−1.05	0.5 M TEAP in DMF	128
$[TcI_2(diars)_2]^{0,1-}$	−0.77	0.5 M TEAP in DMF	128
tr-$[Tc(NO)(H_2O)(NH_3)_4]^{3+,2+}$	0.80	3 M HTFMS	68
$[Tc(NO)(NCS)_5]^{2-,3-}$	0.38		171
$[Tc_2Cl_8]^{2-,3-}$	0.38	10% v/v 12 M HCl in EtOH	172

[a] For comparison purposes all potentials are reported relative to the NHE, regardless of the reference electrode employed experimentally. Corrections relative to the SCE were made by adding 0.242 V to the experimental value

3. *Technetium(V)*

A relatively large number of Tc(V) complexes have been recently prepared and characterized by x-ray diffraction methods. The square pyramidal $[TcOCl_4]^-$ ion had previously been synthesized and characterized on the basis of its infrared and electronic spectra[81, 82], but has only recently been utilized as a versatile synthetic starting material owing to the facile preparation devised by Cotton, Davison and collaborators[83, 84]. The ion has a Tc=O bond distance of 1.610 Å and Tc-Cl distances of 2.31 Å with a slight distortion from square pyramidal geometry evident in the Cl-Tc-Cl angles of 139.2° and 153.7°[83]. The average O-Tc-Cl bond angle is 106.8° and the Tc atom is displaced from the plane of the chlorines by 0.66 Å (toward the oxygen). The complex $[TcOBr_4]^-$ is available by a similar synthetic method[85], while the analogous iodo complex is produced by the ligand exchange reaction between $[TcOCl_4]^-$ and NaI in acetone[86].

Similar square pyramidal structures have been observed for: $[TcO(SCH_2CH_2S)_2]^-$ [87], $[TcO(SCH_2COS)_2]^-$ [88], and $[TcO(SCH_2CH_2O)_2]^-$ [89], which can all be prepared from $[TcOCl_4]^-$ through ligand substitution. In the thiomercaptoacetate complex, the Tc=O bond length is 1.672 Å and the Tc-S lengths are 2.336 Å, and 2.303 Å, with the technetium atom lying 0.791 Å above the plane of the sulfurs[88]. The structure of the tetraphenylarsonium salt of the ethanedithiolate complex exhibits a Tc=O bond length of 1.64 Å and an average Tc-S length of 2.30 Å[87]. Other analogous compounds are those of 1,3-dimercaptopropane, 1,2-dimercaptopropane[87], maleonitriledithiolate[90], and thiooxalate[91]. A dimeric μ-$O[O(Et_2NCS_2)_2Tc]_2$ compound has been prepared by the action of sodium diethylcarbamate on $(nBu_4N)[TcOCl_4]$ in acetone[92]. By analogy to the corresponding rhenium complex, its structure is thought to involve a linear O=Tc-O-Tc=O backbone.

The oxo moiety in these complexes exerts a strong trans effect in both the structural and kinetic senses. In the case of dichloro[hydrotris(1-pyrazolyl)borato]oxotechnetate(V)[93] hexacoordination is forced by the chelating ligand; however, the three pyrazole nitrogens are not equivalent. The Tc-N bond distances are 2.086 Å, 2.088 Å and 2.259 Å, with the latter distance belonging to the nitrogen trans to the oxygen.

The biodistribution of a series of stable species containing diaminodithiolate ligands has been determined in an effort to develop improved agents for kidney function[94]. An initial structure study of one of these compounds, $(Ph_4As)[TcO(ema)]$ confirms the square pyramidal geometry now typical of mono-oxo Tc(V) compounds[95]. The high stability, convenient half-life and rapid urinary excretion of this complex causes it to be highly promising for renal evaluation.

Aqueous solutions of $K_4[Tc(CN)_7]$ allowed to stand in air decompose to yield $K_2[TcO(CN)_5]$[96], which can also be prepared by the action of hot aqueous KCN on freshly formed TcO_2. Hydrolysis of the mono-oxo Tc(V) complex in water yields $K_3[TcO_2(CN)_4]$, which is also possible through the treatment of $[TcO_2(Pyr)_4]ClO_4 \cdot 2\,H_2O$ with aqueous KCN[96]. Reaction of $(nBu_4N)[TcOCl_4]$ with KCN in methanol yields $(nBu_4N)[TcO(OCH_3)(CN)_4]$, while the analogous reaction with NH_4SCN produces microcrystals of $(Ph_4As)(TcO(NCS)_5)$ upon the addition of $(Ph_4As)Cl$[92].

Recent work has shown that technetium(V) complexes with most common nitrogen ligands tend to adopt an octahedral geometry incorporating a trans dioxotechnetium(V) ion rather than the mono-oxo, square pyramidal structure. The synthesis of these complexes with a variety of nitrogen ligands is fairly straightforward beginning from either

$[TcOCl_4]^-$ or $[TcO_4]^-$; however, the latter also requires the presence of a suitable two-electron reductant[97,98]. Stable complexes with chelating ligands are possible in aqueous solution, whereas ions containing monodentate ligands can usually be prepared only in nonaqueous media, such as alcohol or tetrahydrofuran.

Several examples of complexes involving tr-$[O_2Tc]^+$ with nitrogen ligands have now been characterized by x-ray diffraction. In the compound tr-$[O_2(cyclam)Tc]ClO_4 \cdot H_2O$ the Tc=O bond distances are 1.751 Å and the Tc-N lengths are 2.125 Å[99]. The structure of tr-$[TcO_2(Im)_4]Cl \cdot 2\,H_2O$ reveals two crystallographically separate technetium ions with each residing at a center of symmetry. The Tc=O bond distances for the two independent complex ions are 1.73 Å and 1.76 Å. The Tc-N distances are essentially identical between the two ions and are 2.12 Å and 2.17 Å for each pair of imidazoles trans to one another[100].

Both the iodide and chloride salts of the ethylenediamine complex have been characterized. In the iodide salt the complex sits on a center of symmetry and exhibits a $\delta\lambda$ conformation with respect to the ethylenediamine groups. However, the technetium ions in the chloride salt reside in general positions and exhibit both the $\delta\delta$ and $\lambda\lambda$ geometries in equal proportion (see Fig. 1). Average bond distances for these complexes are: 1.746 Å for Tc=O and 2.15 Å for Tc-N. Since a low activation energy separates the various "twistomeric" geometries, only a single proton resonance is observed by NMR at 37 °C. The complex is stable for hours at neutral pH, but releases the ethylenediamine in acid and undergoes a biphasic decomposition with a half-life of approximately 2.5 min at pH 4[97]. Other trans-dioxotechnetium(V) complexes have been prepared with 1,2- and 1,3-diaminopropane[97,98], cyanide[96], pyridine[101], and 1,2-diaminocyclohexane[97].

The relatively longer Tc=O bond distances in the dioxo species can be attributed to mutual competition between the two oxygen atoms for the same metal orbitals. It has also been suggested that the determining factor between mono- and dioxo complexes is the pi-donor ability of the equatorial ligands[99]. The Tc=O stretching frequencies are of considerable use in detecting this group by both infrared and Raman spectroscopy and in distinguishing between Tc=O (930–1020 cm^{-1}) and O=Tc=O (820–880 cm^{-1}).

Technetium complexes of citrate, mannitol and glucoheptonate were evaluated early as renal-imaging agents and their structures were thought to involve chelation of the metal by deprotonated hydroxyls[102]. Compounds of this type have recently been characterized with catechol and ethyleneglycol ligands[95]. A structure determination of $(nBu_4N)[TcO(cat)_2]$ reveals the Tc to be coordinated in a square pyramidal geometry with an average Tc-O (catecholato) distance of 1.954 Å. It is likely that this geometry also holds for the complexes involving ethyleneglycolate and radiopharmaceuticals incorporating carbohydrate ligands[103,104]

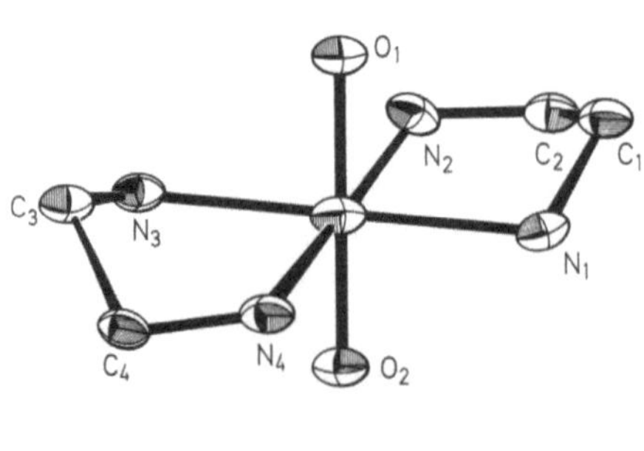

Fig. 1. ORTEP view of tr-$[TcO_2(en)_2]^+$ as the $\lambda\lambda$ form in the chloride salt

DePamphilis and coworkers have undertaken a mechanistic study of the reaction of dithiolate ligands with $[TcOCl_4]^-$ and the analogous catecholato and ethyleneglycolato complexes in methanol[95]. Addition of the dithiolate to the catecholato complex involves a monomeric intermediate apparently containing one sulfur ligand, while reactions with the ethyleneglycolato species proceed to the final product, $[TcO(SCH_2CH_2S)_2]^-$, rapidly and without formation of an observable intermediate. Substitution onto the tetrachloro complex proceeds through a dimer, $[(TcO)_2(SCH_2CH_2S)_3]$, which, surprisingly, contains the technetyl ions in nonequivalent coordination geometries[104]. In this species one technetium is chelated by a single dithiolate and is bridged via the sulfur atoms on two additional dithiolates to the second technetium. The remaining sulfurs of these two dithiolates are coordinated to the second Tc(V). Both metal ions are square pyramidal, but are canted relative to one another by almost 90°. Dimeric products of this type appear to be favored when the dithiolate ligands sterically hinder the approach of additional ligands to the otherwise vacant sixth coordination position on the Tc(V) ions. The structure of some ^{99m}Tc radiopharmaceuticals, such as the hepatobiliary agent containing dihydrothioctic acid, may be quite sensitive to the structure of the incoming ligand and the mechanistic details of the synthesis employed. These workers have also pointed out that ligand substitution onto monomeric oxotechnetium(V) dithiolate complexes probably proceeds by protonation of the thiolate ligands prior to their displacement.

The structure of $[TcO(EDTA)]^-$ is illustrated in Fig. 2 and shows the Tc(V) to be seven-coordinate[106]. Stable, highly water soluble ions such as this are fairly readily eliminated by the kidneys and this structure may be relevant to the renal imaging agents employing EDTA and DTPA.

The molecular structure of $[Tc(DMG)_3SnCl_3(OH)] \cdot 3\ H_2O$ clearly shows that Sn(II) can be incorporated into technetium complexes prepared by stannous reduction of $[TcO_4]^-$ [107]. The Tc and Sn atoms are bridged by an oxygen atom and twice by N-O linkages from different DMG ligands. The tin atom is octahedral while the Tc atom is seven coordinate in a capped trigonal prism. Six imino nitrogens and the bridging oxygen comprise the technetium sphere. The Tc-N distances are all between 2.074 Å and 2.104 Å, while the Tc-O length is 2.030 Å. The Tc-Sn distance of 3.47 Å suggests that there is no intermetallic bonding.

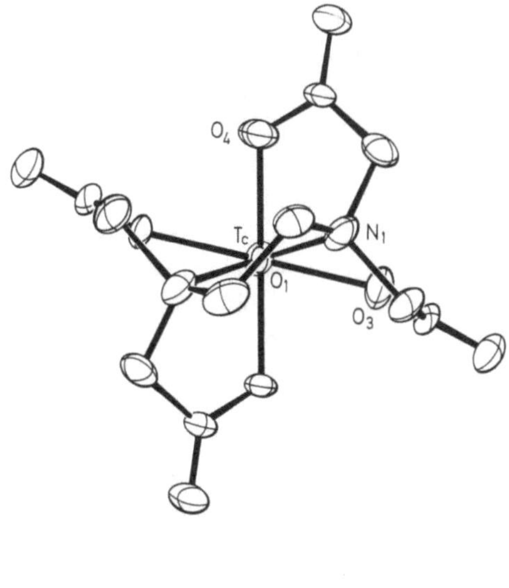

Fig. 2. ORTEP view of $[TcO(EDTA)]^-$ down the crystallographic two-fold axis. Bond lengths are: Tc=O_1, 1.657 Å; Tc-O_3, 2.159 Å; Tc-O_4, 2.028 Å; Tc-N_1, 2.350 Å. The technetium atom is directly in front of O_1 in this view

Oxidation of the octahedral $[Tc(diars)_2Cl_2]^+$ with chlorine yields $[Tc(diars)_2Cl_4]^+$, which exhibits a coordination number of eight with D_{2d} symmetry. This oxidation of Tc(III) to Tc(V) represents the first example of oxidative addition from a six coordinate to an eight coordinate complex[108]. Other Tc(V) complexes include TcF_5, $[TcF_6]^-$, $TcOCl_3$[76], and $TcOBr_3$[75].

4. *Technetium(IV)*

While freshly prepared $[TcO_2 \cdot 2\,H_2O]_n$ is occasionally employed as a starting material[94,96], the hexahalo-dianions, $[TcX_6]^{2-}$ (where X = Cl, Br or I), are more commonly used. The relative stabilities of these complexes are Cl > Br > I[109]. The chloro and bromo complexes are readily synthesized by the reduction of $[TcO_4]^-$ in hot solutions of the corresponding hydrohalic acid[105]. Reduction proceeds through $[TcOX_4]^-$, which can be isolated with bulky cations from cool solutions[83,85]. The iodo complex is prepared from the corresponding chloro or bromo species by evaporation in HI[109]. A crystal structure determination of $(NH_4)_2[TcCl_6]$ shows the complex to be octahedral with the Tc-Cl bond distances of 2.33 Å[110,111]. In aqueous solution these ions readily hydrolyze to form the hydrated dioxide, and are therefore usually handled in strong solutions of the respective hydrohalic acid or in nonaqueous media.

Electrolytic reduction of solutions of $[TcCl_6]^{2-}$ produces a brown deposit, which has been variously suggested to be $[TcCl_5(OH)]^{3-}$ or $[TcCl_5(OH_2)]^{2-}$[112,113]. Anodic or aerobic reoxidation of this species in 1 M HCl regenerates the starting ion[112]. The electronic spectra of the hexahalotechnetium(IV) complexes have been extensively studied[114,115]. The optical electronegativity for Tc(IV) is 2.25 and the Jørgensen "g" and "k" values for estimating ligand field strengths and the magnitude of the nephelauxetic effect are 30 and 0.3, respectively. The crystal structure of the polymeric $(TcCl_4)_n$ has also been reported with Tc-Cl bond distances in the range 2.23 Å through 2.45 Å[116].

The compounds $K_2[Tc(CN)_6]$[75,117], $Tc(Ph_3P)_2Cl_4$, $Tc(Ph_3As)_2Cl_4$[118], $Tc(bipy)Cl_4$ and $[Tc(bipy)_2Cl_2]Cl_2$[119] have all been reported; however, more recent work suggests that the latter is actually a mixture of several bipyridine complexes, some of which are unstable in air[100].

Technetium(IV) complexes with phosphate or phosphonate ligands are of particular interest since these species represent components of an important class of bone-imaging radiopharmaceuticals. The solid state structure of a Tc-MDP complex with the empirical formula $Li(H_2O)_3[Tc(OH)(MDP)] \cdot 1/3\,H_2O$ consists of infinite polymeric chains containing two symmetry related Tc ions bridged by a hydroxyl and each coordinated by two symmetry related MDP ligands[120]. Since it was not possible to resolve the hydrogen atoms in this structure, the metal oxidation state cannot be unequivocally assigned; however, it seems reasonable that the technetium is in the IV or, perhaps, V oxidation state. The technetium distances to the phosphate oxygens are between 1.96 Å and 2.03 Å, while the Tc to non-phosphate oxygen bond is 1.912 Å, which is indicative of a singly rather than doubly bonded Tc-O.

Deutsch has summarized the modes by which diphosphonate ligands bridge technetium ions and has pointed out that the hydroxyl group on the central carbon of HEDP allows this ligand to serve as a mixed bidentate-tridentate bridge between metal ions[16,19]. The tridentate binding mode provided by such ligands may allow the technetium complex to coordinate to Ca(II) on the fastest growing crystal axis of hydroxyapatite. It is very probably this ability of diphosphonates to bridge between metal centers which provides for the initial sorption of these radiopharmaceuticals onto bony tissue. The various components present in clinically used ^{99m}Tc diphosphonate preparations have been separated by HPLC techniques and the biological behavior of the fractions evaluated[121, 122]. A negative correlation exists between the negative "charge density" of the radiopharmaceutical as measured by the logarithm of its chromatographic retention time and its bone uptake.

A series of Tc(III) and Tc(IV) complexes with acetylacetone has been prepared by refluxing $(Ph_3P)_2TcX_4$ or $(Ph_4P)_2TcX_6$ in pentane-2,4-dione[123]. Included in this series are: $(Ph_3P)(acac)TcBr_3$, $(PPh_4)[(TcX_4(acac)]$, $TcX_2(acac)(PPh_3)_2$ and $TcX_2(acac)_2$, where X = Cl, Br. These reactions proceed in a complex manner and, depending on reaction conditions, different products are obtained.

5. *Technetium(III)*

Technetium(III) is one of the more accessible oxidation states and has been shown to complex with a variety of ligands in several coordination geometries. In general, the trivalent state of technetium appears to require the presence of good pi-acceptor ligands such as triphenylphosphine, diarsines, or carbonyl for stabilization.

The compound tr-$[Cl(acac)_2(Ph_3P)Tc]$ can be prepared by direct action of pentane-2,4-dione with tr-$[Tc(PPh_3)_2Cl_4]$[124]. This species exhibits a tetragonal distortion from a regular octahedron and approaches D_{4h} symmetry. The four Tc-O bond distances average 2.01 Å and the two Tc-Cl and Tc-P bond lengths are 2.42 Å and 2.46 Å respectively[125]. If the reaction time is substantially increased, the resulting crystals remain similar in color and rate of appearance; however, the compound crystallizes in a different unit cell with significant changes in the infrared spectrum[125]. Application of the angular overlap method of estimating molecular orbital energies suggests that a greater π-stabilization energy causes the trans configuration to be favored over the cis[124]. It is likely that this is a general result for Tc(II) and Tc(III) complexes with π-delocalizing ligands, since the structures of all bisdiphosphine and bisdiarsine complexes solved to date exhibit this geometry. The complex $[Tc(acac)_3]$ can be prepared by refluxing $[TcO_4]^-$[98] or $TcCl_4(PPh_3)_2$[123] in anhydrous acetylacetone.

The compound mer-$[Cl_3(Me_2PhP)_3Tc]$ is possible by allowing $(NH_4)_2TcCl_6$ to react with dimethylphenylphosphine in absolute ethanol or, alternatively, $(NH_4)TcO_4$ can be mixed with the phosphine in the presence of hydrochloric acid[126]. The technetium-ligand bond distances are: Tc-Cl (trans to P) 2.46 Å; Tc-Cl (trans to Cl) 2.33 Å; Tc-P (trans to Cl) 2.42 Å; Tc-P (trans to P) 2.47 Å. There are two crystallographically independent molecules in the unit cell, which differ only by the orientation of the phenyl rings on the phosphines.

The series of complexes $[TcL_2X_2]^+$ (where L = diphosphine or diarsine ligands, and X = Cl or Br) have been synthesized by allowing the appropriate ligand to react with $[TcX_6]^-$ in acidic ethanol-water mixtures. Pertechnetate at the carrier free level can also be used as the starting material, when phosphine ligands, which serve both as the reductant and chelating agent, are employed[127, 128]. The structure of $[Tc((Me_2P)_2C_2H_2)_2Cl_2]F_3CSO_3$ appears to be representative of this class of complexes and reveals a trans octahedral arrangement of the ligands around the metal ion with Tc-Cl distances of 2.323 Å and Tc-P lengths of 2.436 Å[127].

Crystal structures of the complex tr-$[Cl_2(diars)_2Tc]^+$ have been determined on both the chloride and perchlorate salts[128]. In these compounds, the technetium ion exhibits ideal D_{2h} symmetry with a mean Tc-As distance of 2.512 Å and an average Tc-Cl length of 2.322 Å. The compound $(CO)Cl_3(Me_2PhP)_3Tc$ is seven-coordinate with the carbonyl residing in the center of the three phosphine groups to form a capped octahedral face and the chlorides forming the uncapped face[130]. The metal-ligand bond lengths are: Tc-C, 1.86 Å; Tc-P, 2.44 Å and Tc-Cl, 2.48 Å.

The reaction of $(NH_4)SCN$ in aqueous solution with $(NH_4)_2TcX_6$ (X = Cl or Br) yields $[Tc(NCS)_6]^{2-}$ together with the analogous Tc(III) complex[131]. The Tc(III) ion is octahedral with Tc-N bond distances averaging 2.045 Å. Electrochemical studies in acetonitrile reveal a reversible one-electron transfer between the two oxidation states, so that solutions of $[Tc(NCS)_6]^{3-}$ in air are quickly oxidized to the Tc(IV) form.

Lipophilic Tc(III) complexes with iminodiacetic acid derivatives are readily extracted by the liver and so constitute an important class of hepatobiliary imaging agents[30, 31]. Electrophoretic analysis has shown these complexes to be mononegative and to involve two IDA ligands, which probably hold the metal in an octahedral environment[17, 132]. A linear correlation exists between the biliary clearance of these complexes and their lipophilicity, which was expressed as the natural logarithm of the molecular weight of a complex divided by its charge[17, 132].

6. *Technetium II*

Early work by Fergusson resulted in a series of compounds of the type $[X_2L_2Tc(II)]$, similar to the Tc(III) complexes noted above[118, 119], where L = diphos or diars, X = Cl, Br, or I. The structure of tr-$[Cl_2(diphos)_2Tc]$ and related species have recently been determined[15, 106]. The crystal structure of the analogous complex tr-$[Cl_2(PPh(OEt)_2)_4Tc]$[133] shows a distorted octahedral geometry with Tc-Cl and Tc-P bond distances of 2.41 Å[133, 134]. The series of complexes $TcX_2(PPh(OEt)_2)_4$ reacts with carbon monoxide at atmospheric pressure to produce cis and trans dicarbonyl isomers by replacement of the halides[135].

The compound $(nBu_4N)[Tc(NO)Br_4]$ can be prepared in high yield by the reaction of nitric oxide with freshly precipitated $TcO_2 \cdot xH_2O$ in 4 M HBr[94]. While the bromide ligands of this compound are replaced with some difficulty, it has been possible to use this as a starting material to produce other Tc(II) compounds such as $(nBu_4N)[Tc(NO)Cl_4]$ and $(nBu_4N)_2[Tc(NO)(NCS)_5]$. The complex tr-$[(OH_2)(NO)(NH_3)_4Tc]^{3+}$ is available through oxidation of the corresponding Tc(I) species and exhibits a remarkably acidic pK_a of 2.0[68]. The proposed square pyramidal structure of the bromo complex suggests that the nitrosyl group exerts a fairly strong trans effect[94]. The action of excess NCS^- on $[Tc(NO)Br_4]^-$ allows the isolation of crystalline $[nBu_4N]_2[Tc(NO)(NCS)_5]$, which can subsequently be reduced with hydrazine to yield the analogous Tc(I) species. Both these complexes are thought to involve nitrogen coordination of the thiocyanate.

7. *Technetium I*

The compound tr-$[Tc(NO)(OH_2)(NH_3)_4Cl_2$ is prepared by the reaction of hydroxylamine with $[TcCl_6]^{2-}$ [68]. The Tc-NO bond length in this compound is quite short (1.716 Å) and is indicative of strong d_π-π^* retrodative bonding. Delocalization of electron density away from the metal in this manner undoubtedly strengthens the Tc-NH_3 bonds (2.163 Å), which cannot be substituted even under forcing conditions. This effect also accounts for the relatively low pK_a (7.3) of the coordinated water. The fairly long Tc-OH_2 bond length of 2.168 Å suggests the presence of a trans effect exerted by the nitrosyl.

The compound $(nBu_4N)_3[Tc(NO)(NCS)_4]$ can be synthesized by reduction of the corresponding Tc(II) species to which it can also be reversibly oxidized, while $Tc(NO)Br_2(Pyr)_3$ is available by refluxing $(Me_4N)[Tc(NO)Br_4]$ in pyridine[101].

Reaction of $Tc_2(CO)_{10}$[136, 137] with mesoporphyrin IX dimethylester or meso-tetraphenylporphine produces both mono-metallic and bimetallic porphyrin complexes[138]. When the molar ratio of porphyrin to $Tc_2(CO)_{10}$ is approximately 1:1, a bimetallic compound is formed with each metal atom bonding to three of the four nitrogen atoms on opposite sides of the porphyrin. When the molar ratio is 1:0.6 (porphine to metal carbonyl) a mononuclear complex of the monohydrogenated porphyrin results, which undergoes thermal decomposition to the bimetallic species and dihydrogenporphyrin. The metal-nitrogen bond distances in the bimetallic compound are 2.16 Å and 2.39 Å. The metal-metal distance is 3.101 Å, indicating the absence of an intermetallic bond.

The structure of cis-$[Tc(CO)_2(P(OEt)_2Ph)_4]ClO_4$ shows the metal to be in a distorted octahedral geometry with a mean Tc-P bond distance of 2.41 Å, and Tc-CO bond lengths of 1.90 Å. Other reported Tc(I) compounds have also been stabilized by the presence of strong π-acceptor ligands. These include: $Tc(C_5H_5)(CO)_3$[139], $[Tc(C_5H_5)_2]_2$[140], $[Tc(C_6H_6)_2]^+$ [139], $[Tc(CO)_5]X$ (where X = H^-, Cl^-, Br^-, I^-)[141], $[Tc(CO)_4X]_2$[141], and $(C_6H_5COC_5H_4)Tc(CO)_3$[142]. A compound formulated as $K_5[Tc(CN)_6]$ has been reported by Schwochau to result from the reduction of $[TcO(CN)_5]^{2-}$ with potassium amalgam in the presence of an alkaline KCN solution. This species is easily reoxidized in air back to the starting material[143].

8. *Complexes with Tc-Tc Bonds*

Two dimeric Tc(III) complexes, $[Tc_2(O_2CCMe_3)_4Cl_2]$ and $(nBu_4N)_2[Tc_2Cl_8]$, provide interesting examples of what are thought to be Tc-Tc quadruple bonds. The internuclear distances in these complexes are 2.192 Å[144] and 2.147 Å[94], respectively. In the former complex, the carboxylate groups bridge between the Tc atoms and the chlorides occupy the axial positions[144]. The average Tc-O bond length is 2.032 Å, while the Tc-Cl distance is 2.408 Å. The latter compound is strictly isomorphous with $[nBu_4N]_2[ReCl_8]$ and exhibits D_{4h} symmetry with average Tc-Cl bond distances of 2.33 Å.

Two structures containing the metal in the average oxidation state of 2.5 with a Tc-Tc bond order of 3.5 have also been reported. In $[Tc_2Cl_8]^{3-}$ the chlorides are held in the eclipsed configuration characteristic of complexes containing higher order metal-metal bonds[145]. Despite its lower bond order, the Tc-Tc distance is shorter (2.117 Å) than in either of the afore mentioned complexes – an anomaly sure to be the subject of future theoretical speculation. Owing to the short bridging distance allowed by 2-pyridone, the compound $[Tc_2(2\text{-oxopyridinato})_4Cl]$ contains the shortest Tc-Tc bond (2.095 Å) to date[146]. The chlorides in this compound are axially situated at a distance of 2.679 Å between the technetiums of adjacent units, so as to join them in infinite chains.

In $Tc_2(CO)_{10}$ the octahedral technetium atoms are held together by a single bond of 3.036 Å in length[147]. The equatorial carbonyls of each are arranged in a staggered configuration and there is a significant difference between the equatorial and apical ligands. Since the apical carbonyls are able to enter into a stronger backbonding interaction with the metal, they exhibit a Tc-C bond distance (1.90 Å) that is 0.1 Å shorter than that of the equatorial ligands. In harmony with this, the greater acceptance of the electron density into the π^*-orbital of the apical carbonyls causes their C-O bond lengths to be 0.08 Å longer than in the equatorial carbonyls.

C. Electrochemistry

Since the efficient reduction of very dilute solutions of $[TcO_4]^-$ is required for the production of most ^{99m}Tc radiopharmaceuticals, a fairly large number of polarographic studies have been performed on this ion. However, most of these have yielded inconsistent results. It has recently been pointed out that mercury is capable of reducing pertechnetate, especially in the presence of hydrohalic acids[15] and this may have been a complicating factor in many early studies.

Cyclic voltammetric investigations of the reduction of $[TcO_4]^-$ on platinum[148] and mercury[149] over a broad pH range have helped to clarify a number of earlier polarographic investigations[150–157]. These studies indicate that the ion is reduced by hydrogen atoms on the surface of the platinum electrode producing TcO_2 and Tc_2O_3, while these species are formed by direct charge transfer on mercury electrodes. It is now thought that the initial polarographic reduction wave is due to a diffusion controlled, pH dependent 4 e^- reduction, which is followed by a subsequent 3 e^- process to yield the metal[15, 149, 158]. The reoxidation of the reduced oxides proceeds through intermediate oxidation states, depending on the pH[148]. Potential versus pH diagrams illustrating the thermodynamically preferred reduction products for a range of $[TcO_4]^-$ concentrations have been constructed to conveniently summarize the results of the available polarographic data[149].

Fast-scan cyclic voltammetry studies on $[TcO_4]^-$ in aqueous solution indicate an initial 1 e^- reduction to yield $[TcO_4]^{2-}$. This species is formed only in basic media and exists transiently with a millisecond lifetime[79]. Cyclic voltammetry studies of $Cl_3Tc(PMe_2Ph)_3$ in CH_3CN indicate that chloride ligands can be lost on either oxidation or reduction with the solvent occupying the vacated position[159, 160]. The complex $[Cl_2Tc(CH_3CN)(PMe_2Ph)_3]^+$ has been prepared by these means and its formation verified by electrochemical methods[160]. The reduction potentials of a series of Tc(III) complexes of the type $[TcL_2X_2]^+$, where X = halogen and L = (diphenylphosphino)ethane, -ethylene, -benzene, (diethylphosphino)ethane and (dimethylphosphino)ethane, lie in the range 0.40 V to −0.25 V and correlate in a linear fashion with the energy of the visible charge-transfer bands (460 nm to 600 nm)[127]. Electrochemical studies on hexahalotechnetium(IV) ions reveal these species to undergo rapid ligand substitution upon reduction to Tc(III)[38, 128].

Srivastava and Richards have recently surveyed a broad range of reductants commonly employed in the production of technetium radioscintigraphic agents[15], while Jones et al. have found that dithionite is an excellent reducing agent for the formation of oxotechnetium(V) complexes with dithiolate ligands[35]. Others have adapted this technique to develop facile methods for the synthesis of dioxotechnetium(V) complexes with nitrogen ligands[98, 100].

Since oxygen atoms must be released in the reduction of pertechnetate and most intermediate species, these processes should be facilitated by the presence of acid or other oxygen acceptors. However, cyclic voltammetric studies with complexes such as tr-$[TcO_2en_2]^+$, reveal no reduction waves within the accessible potential range[100]. Owing to the relatively high technetium oxidation state, these complexes should not be difficult to reduce on a thermodynamic basis, so that the lack of electrochemical reduction probably arises from a kinetic effect. It is likely that redox reactions involving oxotechnetium ions proceed most easily by oxygen atom transfer to the reducing agent. Therefore, it is not surprising that the best pertechnetate reductants for synthetic purposes are those capable of reacting by such a mechanism[15, 35, 38, 40].

D. Magnetic Properties

Since numerous oxidation states are readily available to technetium, a primary consideration in the characterization of new compounds and radiopharmaceuticals is the assignment of this quantity. Table 2 summarizes a series of experimental values for the effective magnetic moments of a variety of technetium compounds in a range of oxidation states. Some Tc(VII) compounds such as Tc_2O_7 also exhibit a weak temperature-independent paramagnetism[75]. Spin-orbit coupling is usually a major factor contributing to the magnetic moments of second-row transition metal ions. On this basis the values for octahedral technetium ions in magnetically dilute environments at 25 °C have been calculated to be: 2.0, Tc(II); 3.2, Tc(III); and 3.6, Tc(IV)[161, 162]. However, reference to Table 2 suggests that the observed magnetic moments are often closer to those estimated from spin-only considerations.

While the magnetic moments of the hexahalotechnetium(IV) complexes are consistent with a $^4A_{2g}$ ground state, these complexes do not exhibit EPR signals at room temperature or 77 °C[38, 109]. A substantial field-dependent, temperature-independent paramagnetism is evident for the series of mono-oxo, square pyramidal Tc(V) complexes. This behavior is typical of oxo, d^2 metal ions with local C_{4v} symmetry in this region of the periodic table[163] and should be of use in confirming the presence of Tc(V) in this geometry. Despite their paramagnetism, the NMR spectra of these complexes can be easily acquired.

Owing to the $I = 9/2$ value of the ^{99}Tc nucleus, EPR spectra involving a single electron localized on technetium exhibit a 10 line pattern[68, 94]. Spectra of this type reported for $[Tc(NO)X_4]^-$ (X = Cl and Br) are consistent with a $^2B_2$ ground state and yield $g_\perp$ values of 2.345 and 2.055 and $g_\parallel$ values of 1.915 and 2.129 for the chloro and bromo complexes, respectively[94]. This nucleus is also NMR active and $[^{99}TcO_4]^-$ exhibits a sharp resonance at 90.06 MHz and 400 KG with $T_1 = 0.13$ s and $T_2 = 0.10$ s[164].

Table 2. Effective magnetic moments of representative technetium complexes

Compound	Oxidation State	Magnetic Moment (BM)	Ref.
$(Me_4N)_2[TcO_4]$	VI	1.6	78, 173
$(Ph_4As)[TcO(SCH_2CH_2S)_2]$	V	2.11[a], 1.16[b], 0.82[c]	163
$(Ph_4As)[TcO(SCH_2CH(CH_3)S)_2]$	V	2.60[a], 1.00[c]	163
$(Ph_4As)[TcO(SCH_2CH_2CH_2S)_2]$	V	2.76[a], 1.07[c]	163
$(nBu_4N)[TcO(SCH_2COS)_2]$	V	1.50[b], 1.19[c]	91
$(nBu_4N)[TcO(SOCCOS)_2]$	V	0.34[b], 0.17[c]	91
$(nBu_4N)[TcO(tdt)_2]$	V	0.42[b], 0[c]	91
$(Ph_4As)[TcO(NCS)_5]$	V	1.0	92
K_2TcCl_6	IV	4.05	109
K_2TcBr_6	IV	3.94	109
K_2TcI_6	IV	4.14	109.
$(diphosH_2)TcCl_6$	IV	3.74	118
$(diphosH_2)TcBr_6$	IV	4.02	118
$TcCl_4(PMe_2Ph)_2$	IV	3.6	174
$TcBr_4(PMe_2Ph)_2$	IV	3.8	174
$TcCl_4(PPh_3)_2$	IV	3.7	174
$TcBr_4(PEt_2Ph)_2$	IV	3.6	174
$TcCl_4(Ph_3As)_2$	IV	3.62	119
$[TcCl_2(bipy)_2]Cl_2$	IV	3.7	119
$TcCl_4(bipy)$	IV	3.48	119
$TcBr_4(PPh_3)_2$	IV	3.97	118
$TcCl_3(PMe_2Ph)_3$	III	2.8	175
$TcBr_3(PEt_2Ph)_3$	III	2.8	175
$[TcCl_2(OEt_2PhP)_4]ClO_4$	III	2.7	134
$[Tc(diphos)_2Cl_2]Cl$	III	2.7	118
$[Tc(diphos)_2Br_2]Br$	III	3.04	118
$[TcBr_2(P(OEt)_2Ph)_4]ClO_4$	III	2.5	135
$[TcI_2(P(OEt)_2Ph)_4]I$	III	2.3	135
$[TcI_2(P(OEt)_2Ph)_4]ClO_4$	III	2.4	135
$[TcCl_2(P(OEt)_2Ph)_4]ClO_4$	III	2.6	135
$(nBu_4N)_3Tc(NCS)_6$	III	3.2	92
$(NH_4)_3Tc_2Cl_8 \cdot 2\,H_2O$	II, III	3.40	176
$Tc(diphos)_2Cl_2$	II	2.05	118
$Tc(diphos)_2Br_2$	II	2.28	118
$(nBu_4N)Tc(NO)Br_4$	II	1.88	94
$(nBu_4N)Tc(NO)Cl_4$	II	1.90	94

[a] measured at 1910 G
[b] measured at 7250 G
[c] measured at 14000 G

E. Prospective Developments

Recent advances in micro and submicro analytical techniques are now beginning to facilitate the characterization of new technetium radiopharmaceuticals. The technique of field desorption mass spectroscopy has been applied to a series of Tc(V) compounds and offers considerable promise in assigning the molecular weight and structural components of new complexes[165]. The number of products and relative electrophoretic mobilities of technetium complexes formed by ligand exchange reactions can be used to establish the charge and ligand stoichiometry of radiopharmaceuticals at the carrier-free level[17, 133]. High pressure liquid chromatography will see increasing use in determining the constituents of diagnostic ^{99m}Tc preparations and their most active components as well as in the purification of new technetium complexes. Both anionic and cationic complexes can be chromatographed by reverse-phase ion-pairing methods using ammonium or alkylammonium salts of short-chain carboxylic acids in aqueous eluants[166]. Addition of up to 10% of organic solvents, such as acetonitrile, ethylacetate or dioxane, often improves the chromatographic characteristics[121, 141]. Correlations between biodistribution and chromatographic retention times may also be of value in predicting the efficacy of new complexes[122].

Current work on the development of ^{99m}Tc radiopharmaceuticals focuses on improving the specificity of agents for individual organs and in devising new compounds to image tissues or physiological functions not yet adequately handled by existing methods. There is particlar interest in the visualization of heart and circulatory abnormalities, especially infarcts and thromboses, as well as pancreatic processes and tumors. Deutsch has pointed out that, due to the requirement of functioning muscles for K^+, heart tissue may concentrate monopositive technetium complexes. Despite the obvious differences between bulky complex ions and K^+, initial investigations have shown that complexes of the type $[X_2L_2Tc]^+$, where X = Cl or Br and L = diarsine or diphosphine ligands, may yield excellent heart-imaging agents[127, 167].

Technetium(V) compounds are likely to constitute a substantial portion of new clinical diagnostic agents, since a range of these complexes are easily prepared from pertechnetate, and several types of oxotechnetium(V) complexes undergo facile ligand substitution. In particular, trans-dioxotechnetium(V) ions with nitrogen ligands may lend themselves well to tumor imaging, since they have the possibility of localizing in neoplasms by either the reduction or hydrolysis mechanisms. Metal ion reduction and subsequent substitution of the oxygen atoms should be more probable on the interior of hypoxic tumors and so cause the isotope to tend to concentrate inside the neoplasm. The lower pH of tumor tissue should also facilitate the hydrolysis of such complexes as tr-$[TcO_2(en)_2]^+$, so as to deposit insoluble ^{99m}Tc products within the tumor.

The tetrahalonitrosyl compounds of both Tc(I) and Tc(II) may provide versatile starting materials for a range of nitrosyltechnetium complexes with a variety of ligands, especially those involving nitrogen coordination. Orvig and coworkers have already pointed out the correspondence between reaction products involving $[Tc(NO)Br_4]^-$ and EDTA or DTPA and the active components of the renal imaging agents with these ligands[94].

Because of its excellent tumor-localizing properties, radiolabelling of the anticancer antibiotic bleomycin has long been of interest. Despite its requirement for iron and strong complexing ability for copper[168, 169], most attempts at direct complexation of bleomycin with ^{99m}Tc have resulted in unstable products[6]. One method of circumventing this problem is to attach a strong technetium chelating agent to the molecule[47]; however, such modification may result in decreased tumor uptake. A better means may be to incorporate either oxotechnetium(V) or nitrosyltechnetium(I) or (II), which possess a substantial affinity for nitrogen ligands and should fit nicely into the square pyramidal coordination shown by the bleomycin copper chelate[168, 169]. The affinity of these ions for nitrogen ligands should also help extend the range of technetium radiopharmaceuticals utilizing biologically important molecules such as amino acids, nucleotides and their polymers.

Finally, while the new synthetic methods for technetium compounds in the oxidation states I, II and V will undoubtedly play an important role in the development of improved radioscintigraphic agents, there remains a need to understand the chemical mechanisms involved in these syntheses, particularly with regard to ligand substitution reactions and atom transfer redox processes. Mechanistic studies to complement the increasing body of structural knowledge is essential to the further development of technetium radiopharmaceuticals.

Acknowledgements. We are especially grateful to E. Deutsch (University of Cincinnatti), S. C. Srivastava and A. Packard (Brookhaven National Laboratories), and B. DePamphilis (Massachusetts Institute of Technology), who have made the results of experimental work available to us prior to publication. Work in our laboratory was supported by the American Cancer Society, Massachusetts Division and PHS Grant CA24344.

References

1. Perrier, C., Segre, E.: J. Chem. Phys. *5,* 712 (1937)
2. Perrier, C., Segre, E.: Nature *24,* 159 (1947)
3. Richards, P., Tucker, W. D., Srivastava, S. C.: Int. J. Appl. Radiat. Isot., in print (1981)
4. Kotegov, K. V., Pavlov, O. N., Shudov, V. P.: Adv. Inorg. Radiochem. *11,* 1 (1968)
5. Anders, E.: The Radiochemistry of Tc, NAS-NS 3021 Government Publication, USA
6. Yokoyama, A.: Metal Ions in Biol. Syst. *10,* 375 (1980)
7. Jones, S. E., Salmon, S. E.: Semin. Nucl. Med. *6,* 331 (1976)
8. McCready, V. R.: Br. Med. Bull. *36,* 209 (1980)
9. Silberstein, E. B.: Am. J. Med. *60,* 230 (1976)
10. Nishiyama, H.: Radiopharmaceuticals for thyroid imaging, in: Radiopharmaceuticals II (ed. Sorenson, J. A.), p. 655, New York, Soc. of Nuclear Medicine 1979
11. Erjavec, M. et al.: J. Nucl. Med. *18,* 346 (1977)
12. Golday, D. L., Ash, J.: Radiology *117,* 93 (1975)
13. Turner, D. A. et al.: J. Nucl. Med. *18,* 258 (1977)
14. Heindel, N. D. et al.: The Chemistry of Radiopharmaceuticals, p. v, Masson Publishing, Inc., New York, 1978
15. Srivastava, S. C., Richards, P.: Radiotracers for Medical Applications, CRC Press, Boca Raton, FL, 1981
16. Deutsch, E.: Inorganic radiopharmaceuticals, in: Radiopharmaceuticals II (ed. Sorenson, J. A.), p. 129, New York, Soc. of Nuclear Medicine, 1979
17. Marzilli, L. G., Dannals, R. F., Burns, H. D.: ^{99m}Tc radiopharmaceuticals: uses, problems, potential applications in nuclear medicine, in: Inorganic Chemistry in Biology and Medicine (ed. Martell, A. E.), p. 91, ACS Symp. Series #140, Washington, Amer. Chem. Soc. 1980
18. Francis, M. D. et al.: Imaging the skeletal system, in: Radiopharmaceuticals II (ed. Sorenson, J. A.), p. 603, New York, Soc. of Nuclear Medicine 1979

19. Deutsch, E., Barnett, B. L.: Synthetic a. structural aspects of Tc chemistry as related to nuclear medicine, in: Inorganic Chem. in Biology a. Medicine (ed. Martell, A. E.), p. 103, ACS Symp. Series #140, Washington, Amer. Chem. Soc. 1980
20. Libson, K., Deutsch, E., Barnett, B. L.: J. Am. Chem. Soc. *102,* 2476 (1980)
21. Davis, M. A., Jones, A. G.: Semin. Nucl. Med. *6,* 19 (1976)
22. Subramanian, G., McAfee, J., Blair, R.: J. Nucl. Med. *16,* 744 (1975)
23. Dewanjee, M. K., Brueggemann, P., Wahner, H.: Affinity constants of $^{99m}TcO_4^-$ and Tc-chelates with human serum albumin, in: Radiopharmaceuticals II (ed. Sorenson, J. A.), p. 435, New York, Soc. of Nuclear Medicine 1979
24. Siegel, J. A., Deutsch, E.: Ann. Rep. in Inorg. a. Gen. Synth.-(1975), Academic Press, New York, 1976
25. Ege, N. G., Warbick-Cerone, A., Bronskill, M. J.: Radionuclide lymphoscintigraphy- -an update, in: Radiopharmaceuticals II (ed. Sorenson, J. A.), p. 241, New York, Soc. of Nuclear Medicine 1979
26. Winchell, H. S.: Radiopharmaceuticals in evaluation of kidneys, ibid., p. 459, New York, Soc. of Nuclear Medicine 1979
27. Poe, N. D.: Semin. Nucl. Med. *7,* 7 (1977)
28. Lutzker, L. G., Alvi, A.: ibid. *6,* 83 (1976)
29. Jones, A. G. et al.: J. Nucl. Med. *18,* 680 (1977)
30. Loberg, M. D., Porter, D. W.: Review and current status of hepatobiliary imaging agents, in: Radiopharmaceuticals II (ed. Sorenson, J. A.), p. 519, New York, Soc. of Nuclear Medicine 1979
31. Loberg, M. D., Fields, A. T.: Int. J. Appl. Radiat. Isot. *29,* 167 (1978)
32. Srivastava, S. C. et al.: ibid. *28,* 83 (1977)
33. Steigman, J., Meinken, G., Richards, P.: ibid. *29,* 653 (1978)
34. Eckelman, W. C., Levenson, S. M.: Radiopharmaceutical chem. of Tc and I, in: Textbook of Nuclear Medicine: Basic Science (eds. Rocha and Harbert), p. 192, Philadelphia, Lea and Febiger, 1978
35. Jones, A. G. et al.: J. Nucl. Med. *21,* 279 (1980)
36. Chervu, L. R. et al.: ibid. *18,* 62 (1977)
37. Galatenau, I., Bratu, C., Petride, A.: Radiochem. Radioanal. Lett. *28,* 95 (1977)
38. Trop, H. S. et al.: J. Inorg. Nucl. Chem. *41,* 271 (1979)
39. Shukla, S. K.: J. Chrom. *151,* 517 (1978)
40. Fritzberg, A. R., Lyster, D. M., Dolphin, D. H.: J. Nucl. Med. *18,* 553 (1977)
41. Russell, C. D., Majerik, J.: Int. J. Appl. Radiat. Isot. *29,* 109 (1978)
42. Russell, C. D.: ibid. *28,* 241 (1977)
43. Jones, A. G.: J. Nucl. Med. *21,* 279 (1980)
44. Welch, M. J., Moerlein, S.: Radiolabeled compounds of biomedical interest containing Ga and In, in: Inorganic Chem. in Biology a. Medicine, (ed. Martell, A. E.), p. 121, ACS Symp. Series #140, Washington, Amer. Chem. Soc. 1980
45. Lilien, D. L.: J. Am. Med. Assoc. *230,* 735 (1974)
46. Krohn, K. A. et al.: J. Nucl. Med. *18,* 276 (1977)
47. Goodwin, D. A., Mears, C. F.: Semin. Nucl. Chem. *6,* 389 (1976)
48. Roberts, J. G. et al.: Lancet *1,* 237 (1976)
49. Lundell, G. et al.: Acta Radiol. Ther. *14,* 333 (1975)
50. Osmond, J. D., Prendergrass, H. P., Potsaid, M. S.: Am. J. Roentgenol. Radiat. Ther. Nucl. Med. *125,* 9172 (1975)
51. Rose, C.: Int. J. Nucl. Med. Biol., 142 (1975)
52. Richman, L. S. et al.: Am. J. Roentgenol. Radiat. Ther. Nucl. Med. *124,* 577 (1975)
53. Winchell, H. S.: Semin. Nucl. Med. *6,* 371 (1976)
54. Thews, G., Vaupel, P.: Adv. Exp. Med. Biol. *75,* 537 (1976)
55. Gullino, P. M.: ibid. *75,* 521 (1976)
56. Vaupel, P.: Microvasc. Res. *13,* 399 (1977)
57. Shapot, V. S.: Adv. Exp. Med. Biol. *75,* 581 (1976)
58. Nakamura, W., Hosoda, S.: Biochim. Biophys. Acta *158,* 212 (1968)
59. Tannock, I.: Adv. Exp. Med. Biol. *75,* 597 (1976)
60. Gullino, P., Yi, P. N., Grantham, F. H.: J. Natl. Canc. Inst. *60,* 835 (1978)

61. Grote, J., Susskind, R., Vaupel, P.: Pflugers Arch. *372,* 37 (1977)
62. Cherry, J. M. et al.: Biochim. Biophys. Acta *634,* 11 (1981)
63. Weinhous, S.: Z. Krebsforsch. *87,* 115 (1976)
64. Cater, D. B., Phillips, A. F.: Nature *174,* 121 (1954)
65. Clarke, M. J.: Met. Ion. in Biol. Syst. *11,* 231 (1980)
66. Clarke, M. J.: Potential of Ru in anti-cancer pharmaceuticals, in: Inorganic Chem. in Biology a. Medicine (ed. Martell, A. E.), p. 157, ACS Symp. Series # 140, Washington, Amer. Chem. Soc. 1980
67. Clarke, M. J.: Rev. Inorg. Chem. *2,* 27 (1980)
68. Armstrong, R. A., Taube, H.: Inorg. Chem. *15,* 1904 (1976)
69. Krebbs, B.: Z. Anorg. Allg. Chem. *380,* 145 (1971)
70. Schwochau, V. K.: Z. Naturforsch. *17a,* 630 (1962)
71. McDonald, B. J., Tyson, G. J.: Acta Cryst. *15,* 87 (1962)
72. Faggiani, R., Lock, C. J. L., Poce, J.: ibid. *B36,* 231 (1980)
73. Cotton, F. A., Wilkinson, G.: Advanced Inorganic Chem., p. 887, Wiley-Interscience, New York, 1980[4]
74. Selig, H., Malm, G.: J. Inorg. Nucl. Chem. *25,* 349 (1963)
75. Peacock, R. D.: The Chemistry of Technetium and Rhenium, Elsevier, London 1966
76. Guest, A., Howard-Lock, H. E., Lock, C. J. L.: J. Mol. Spectrosc. *43,* 273 (1972)
77. Guest, A., Lock, C. J. L.: Can. J. Chem. *50,* 1807 (1972)
78. Schwochau, K. et al.: Angew. Chem. *86,* 350 (1974)
79. Deutsch, E. et al.: J. Chem. Soc. Chem. Comm., 1038 (1978)
80. Edwards, A. J., Jones, G. R., Sills, R. J. C.: J. Chem. Soc. *A,* 2521 (1970)
81. Jezowska-Trzebiatowska, B., Baluka, M.: Bull. Acad. Polon. Sci., Ser. Sci. Chim. *13,* 1 (1975)
82. Baluka, M., Hanuza, J., Jezowska-Trzebiatowska, B.: ibid. *20,* 271 (1971)
83. Cotton, F. A. et al.: Inorg. Chem. *18,* 3024 (1979)
84. Thomas, R. W. et al.: ibid. *19,* 2840 (1980)
85. Preetz, W., Peters, G.: Z. Naturforsch. *35b,* 1355 (1980)
86. Peters, G., Preetz, W.: ibid. *36b,* 138 (1981)
87. Smith, J. E., Byrne, E. F., Cotton, F. A.: J. Am. Chem. Soc. *100,* 5571 (1978)
88. DePamphilis, B. V. et al.: ibid. *100,* 5570 (1978)
89. Jones, A. G., DePamphilis, B. V., Davison, A.: Inorg. Chem. *20,* 1617 (1981)
90. Spies, H., Johannsen, B.: Inorg. Chim. Acta *33,* L113 (1979)
91. Davison, A. et al.: Inorg. Chem. *19,* 1988 (1980)
92. Trop, H. S.: Ph. D. Thesis, Massachusetts Inst. of Techn., 1979
93. Thomas, R. W. et al.: J. Am. Chem. Soc. *101,* 4581 (1979)
94. Orvig, C.: Ph. D. Thesis, Massachusetts Inst. of Techn., 1981
95. DePamphilis, B. V.: Ph. D. Thesis, Massachusetts Inst. of Techn., 1981
96. Trop, H. S., Jones, A. G., Davison, A.: Inorg. Chem. *19,* 1993 (1980)
97. Kastner, M. E., Lindsay, M. J., Clarke, M. J.: ibid. (1981) in press
98. Packard, A. B. et al.: J. Nucl. Med. *22,* 70 (1981)
99. Zuckman, S. A. et al.: Inorg. Chem. *20,* 2386 (1981)
100. Clarke, M. J., Kastner, M. E., Fackler, P. H.: unpublished data, 1980
101. Kuzina, A. F., Oblova, A. A., Spitsyn, V. I.: Zh. Neorg. Khim. *17,* 2630 (1972)
102. de Kieviet, W.: Tc-glucoheptonate, chemical structure and tissue distribution, in: Third Int. Symp. on Radiopharmaceutical Chem. (ed. Welch, M. J.), p. 136, St. Louis 1980
103. Russell, C. D., Speiser, A. G.: J. Nucl. Med. *21,* 1086 (1980)
104. Richards, P., Steigman, J.: in, Radiopharmaceuticals (ed. Subramanian, G. et al.), p. 23, New York, Soc. of Nuclear Medicine 1975
105. Davison, A. et al.: unpublished results, 1981
106. Deutsch, E. et al.: unpublished work 1981
107. Deutsch, E. et al.: Proc. Natl. Acad. Sci. USA *73,* 4287 (1976)
108. Glavan, K. A. et al.: J. Am. Chem. Soc. *102,* 2103 (1980)
109. Dalziel, J. et al.: J. Chem. Soc., 4012 (1958)
110. Elder, M., Estes, G. W., Deutsch, E.: Acta Crystallogr., Sect. B *B35,* 136
111. Elder, R. C. et al.: J. Chem. Soc. *A,* 1423 (1967)
112. Muenze, R.: Isotopenpraxis *11,* 29 (1975)

113. Kawashima, M., Koyama, M., Fujinaga, T.: J. Inorg. Nucl. Chem. *38,* 819 (1976)
114. Jørgensen, C. K., Schwochau, K.: Z. Naturforsch. *20 A,* 65 (1965)
115. Jørgensen, C. K.: Oxidation Numbers and Oxidation States, Springer, New York, 1969
116. Elder, M., Penfold, B. R.: Inorg. Chem. *5,* 1197 (1966)
117. Brown, D., Colton, R.: J. Chem. Soc., 714 (1964)
118. Fergusson, J. E., Hickford, J. H.: Aust. J. Chem. *23,* 453 (1970)
119. Fergusson, J. E., Hickford, J. H.: J. Inorg. Nucl. Chem. *28,* 2293 (1966)
120. Libson, K., Deutsch, E., Barnett, B. L.: J. Am. Chem. Soc. *102,* 2476 (1980)
121. Srivastava, S. C. et al.: J. Nucl. Med. *22,* 69 (1981)
122. Pinkerton, T. C. et al.: Intl. J. Appl. Radiat. Isot., in press, 1982
123. Mazzi, U., Roncari, E., Bandoli, G.: Trans. Met. Chem. *4,* 151 (1979)
124. Bandoli, G., Clemente, D. A., Mazzi, U.: J. Chem. Soc. Dalton Trans. *19,* 1837 (1977)
125. Bandoli, G. et al.: Acta Crystallogr., Sect. B *B 34,* 3559 (1978)
126. Bandoli, G., Clemente, D. A., Mazzi, U.: J. Chem. Soc. Dalton Trans. *2,* 125 (1976)
127. Deutsch, E. et al.: Science, in press, 1981
128. Hurst, R. W., Heineman, W. R., Deutsch, E.: Inorg. Chem., in press, 1981
129. Elder, R. C. et al.: Acta Crystallogr., Sect. B *B 36,* 1662 (1980)
130. Bandoli, G., Clemente, C. A., Mazzi, U.: J. Chem. Soc. Dalton Trans., 373 (1978)
131. Trop, H. S. et al.: Inorg. Chem. *19,* 1105 (1980)
132. Burns, H. D. et al.: in, The Chemistry of Radiopharmaceuticals (eds. Heindel, G., Burns, H. D.), p. 269, New York, Masson Publishing 1978
133. Bandoli, G. et al.: Cryst. Struct. Commun. *3,* 293 (1974)
134. Mazzi, U. et al.: Inorg. Chem. *16,* 1042 (1977)
135. Biagini Cingi, M. C. et al.: Inorg. Chim. Acta *13,* 47 (1975)
136. Hileman, J. C., Huggins, D. K., Kaesz, H. D.: J. Am. Chem. Soc. *83,* 2953 (1961)
137. Hieber, W., Herget, C.: Angew. Chem. *73,* 579 (1961)
138. Tsutsui, M. et al.: J. Am. Chem. Soc. *97,* 3952 (1975)
139. Palm, C., Fischer, E. O., Baumgartner, F.: Tetr. Lett., 253 (1962)
140. Huggins, D. K., Kaesz, H. D.: J. Am. Chem. Soc. *83,* 4744 (1961)
141. Hileman, J. C., Huggins, D. K., Kaesz, H. D.: Inorg. Chem. *1,* 933 (1962)
142. Fischer, E. O., Fellman, W.: J. Organomet. Chem. *1,* 191 (1963)
143. Herr, W., Schwochau, K.: Angew. Chem. *73,* 492 (1961)
144. Cotton, F. A., Gage, L. D.: Nouv. J. Chim. *1,* 441 (1977)
145. Cotton, F. A., Shive, L. W.: Inorg. Chem. *14,* 2032 (1975)
146. Cotton, F. A., Fanwick, P. E., Gage, L. D.: J. Am. Chem. Soc. *102,* 1570 (1980)
147. Bailey, M. F., Dahl, L. F.: Inorg. Chem. *4,* 1140 (1965)
148. Mazzochin, G. et al.: Inorg. Chim. Acta *9,* 263 (1974)
149. Grassi, J., Devynck, J., Tremillion, B.: ibid. *107,* 47 (1979)
150. Zhdanov, S. I., Kuzina, A. F., Spitsyn, V. I.: Russ. J. Inorg. Chem. *15,* 803, 662 (1970)
151. Astheimer, L., Schwochau, K.: J. Electroanal. Chem., *14,* 161 and 240 (1967)
152. Terry, A. A., Zittel, H. E.: Anal. Chem. *35,* 615 (1963).
153. Rulfs, C. L., Pacer, R., Anderson, A.: J. Electroanal. Chem. *15,* 61 (1967)
154. Salaria, G. B., Rulfs, C. I., Elving, P. J.: Anal. Chem. *36,* 156 (1964)
155. Love, D. L., Greendale, A. E.: ibid. *32,* 780 (1960)
156. Colton, R. et al.: J. Chem. Soc., 71 (1960)
157. Miller, H. H., Kelley, M. T., Thomason, P. P.: Adv. Polarog. *11,* 716 (1960)
158. Grassi, J. et al.: J. Electroanal. Chem. *88,* 97 (1978)
159. Mazzochin, G. A., Seeber, R., Roncari, I. M.: Inorg. Chim. Acta *29,* 5 (1978)
160. Mazzochin, G. A. et al.: ibid. *29,* 1 (1978)
161. Earnshaw, A.: Introduction to Magnetochemistry, Academic Press, New York, p. 63 (1968)
162. Figgis, B. N., Lewis, J.: Prog. Inorg. Chem. *6,* 1 (1964)
163. Byrne, E. F., Smith, J. E.: Inorg. Chem. *18,* 1832 (1979)
164. Hawkes, G. E., Buckingham, M. J., Thornback, J. R.: Inorg. Chim. Acta *56,* L 41 (1981)
165. Wilson, B. W. et al.: Anal. Lett. *12,* 303 (1979)
166. Clarke, M. J. et al.: Anal. Biochem., in press, 1982
167. Deutsch, E. et al.: J. Nucl. Med. *22,* 897 (1981)
168. Dabrowiak, J. C.: Met. Ions in Biol. Syst. *11,* 305 (1980)

169. Umezawa, H., Takita, T.: Structure and Bonding *40,* 73 (1981)
170. Astheimer, L., Schwochau, K.: J. Inorg. Nucl. Chem. *38,* 1131 (1976)
171. Orvig, C., Davison, A., Jones, A. G.: The preparation a. characterization of some nitrosyl complexes of ^{99m}Tc, in: Third Int. Symp. on Radiopharmaceutical Chem. (ed. Welch, M. J.), p. 148, St. Louis, 1980
172. Cotton, F. A., Pederson, E.: Inorg. Chem. *14,* 383 (1975)
173. Astheimer, L. et al.: J. Chem. Phys. *63,* 1988 (1975)
174. Mazzi, U., De Paoli, G., Traverso, O.: J. Inorg. Nucl. Chem. *39,* 1090 (1977)
175. Mazzi, U. et al.: ibid. *38,* 721 (1976)
176. Bratton, W. K., Cotton, F. A.: Inorg. Chem. *9,* 789 (1970)
177. Burgi, H. B., Anderegg, G., Blauenstein, P.: Inorg. Chem. *20*, 3829 (1981)
178. Michels, G. D., Svec, H. J.: Inorg. Chem. *20*, 3445 (1981)

Note Added in Proof

The molecular structure of the dimeric compound $(\mu\text{-O})_2[H_2EDTA)Tc(IV)]_2$ has recently been reported and exhibits near C_{2h} symmetry[177]. Each Tc(IV) is hexacoordinated and is chelated by a single H_2EDTA. The Tc-Tc distance is short at 2.33 A and indicates at least a single intermetallic bond. The heteronuclear analogs of $Tc_2(CO)_{10}$ containing Tc and either Mn or Re have also been characterized[178].

The Chemistry of Lanthanide Ions in Solution and in Biological Systems

Robert Joseph P. Williams

Inorganic Chemistry Laboratory, Oxford University, South Parks Road, Oxford OX1 3QR, England

An examination of the structure of lanthanide(III) ion complexes in solution using nuclear magnetic resonance is made in considerable detail. The importance of hydration in both the inner and outer spheres is stressed. The structural data are then used in an attempt to understand the stability constants and the rates of exchange of bound ligands. Fluctional properties are also analysed. Finally, the importance of the structures, the stability constants and the kinetic properties are related to the effects of Ln(III) ions, and through them Ca(II) ions, in biological systems.

Structure and Bonding 50

A. Introduction

The chemistry of the fifteen elements from lanthanum to lutecium has been studied in great depth over the last thirty years[1,2]. It was the advance of separation techniques developed in the atomic energy programme which made a decisive difference to our ability to study the elements individually. In this article I shall direct attention to a variety of problems which have arisen concerning the study and use of the lanthanide cations mostly in aqueous solution and mainly in oxidation state III. The first problem I shall attack is that of the structures of their complex ions. Secondly I shall examine the stability constants of these complexes. I shall use nuclear magnetic resonance spectroscopy (NMR) to obtain information about the solution structures of the complexes and their kinetics. Finally I shall pass on to the use of this knowledge in the study of biological problems[3,4,5].

There are two outstanding reasons for the study of the Ln(III) cations as probes of biological action.

(1) They are the simplest cations which have useful physical properties and a chemistry closely allied to that of Ca(II).

(2) Together they form a series which allows, through their physical properties, a knowledge of the interaction, of a (purely) electrostatic nature, between cations and water or organic ligands. No other series of cations offers a parallel advantage either within chemistry or biology. We believe that the action of ions in biology has been refined below the level of discrimination of 0.01 nm. Certainly Sr(II) does not substitute exactly for Ca(II) and often it will not substitute at all. Thus the difference between Ca(II) and Sr(II) can be a precipitate one. It is this that we have to understand and a detailed study of the chelate chemistry of the series of Ln(III) cations is the one way in which we can approach the problem in a semi-empirical way. The major tool for the examination of the structures of Ln(III) complexes in solution is NMR.

B. NMR Methods

The use of the lanthanides ions as NMR reagents was developed in two ways. Hinckley[6] saw the advantages provided by the local high magnetic field of an organic complex of a lanthanide ion in resolving NMR spectra of organic molecules in organic solvents in association with the Ln(III) complexes of the acetylacetone type. D. H. Williams developed the use of these reagents within organic chemistry[7]. At the same time arising from the use of cobalt porphyrin as a structure (shift) probe[8], R. J. P. Williams and his coworkers[9] described the possibility of using Ln(III) cations in aqueous solution to study the structures of Ln(III) complexes. In this application large resonance shifts were not the major concern. In fact both the organic and inorganic shift probes can be used in exactly the same ways – i.e. just to spread out a spectrum or, after assignment analysis, to determine structure. Simultaneously relaxation probes were also developed for structural analysis[8,9]. I give now a brief outline of the principles, see Ref. 10, 11, 12, 13 for details and many additional examples.

I. Shift Probes

All NMR spectra of molecules are averaged over ensembles of conformations. NMR is a relatively slow spectroscopic technique. The average is not just over a variety of vibrational and rotational states but includes conformational changes which occur within about 10^{-5} s. When changes are slower than about 10^{-2} s the different structures give rise to separate NMR transitions which do not contribute to the ensemble. In the intermediate exchange rate conditions lines broaden and shift together and they may be seen to coalesce as they approach the fast exchange situation. The extremes of slow and fast exchange are the easiest to analyse and control of temperature usually allows one or the other condition to be satisfied. Insofar that Ln(III) ions produce very large shifts exchange must be fast if lines are not to be so broadened that fine structure on lines is lost. Intermediate exchange conditions are situations in which use of Ln(III) reagents as simple shift reagents is virtually impossible due to extreme line broadening but much knowledge about rates of exchange can be gleaned. It is sometimes useful to abandon the free cations and to use a variety of very robust Ln(III) cation complexes in water, e.g. Ln(III)EDTA, much as Ln(III) complexes are used in organic media since to these complexes the binding of additional organic ligands is weak and usually the kinetic conditions conform to the fast exchange NMR condition. I describe some useful complexes later in the article. Very little advantages has been taken of this chemical control over the lanthanide reagents in aqueous solutions at the time of writing.

Let us assume that shifted resonances can be easily followed in a titration of Ln(III), or a Ln(III) complex, with an organic molecule. Let us also assume that we can prove that only a single complex of 1 : 1 stoichieometry is formed. We can now analyse the shifts on proton, phosphorus or carbon resonance lines by making the reasonable assumption that there is no free electron "contact" shift except on coordinated atoms. The shifts are then dipolar. Contact shifts will be considered again later.

The dipolar through-space shift due to a paramagnetic ion is given by the equation

$$\delta_p = D' \cdot \frac{(3\cos^2\theta - 1)}{r^3} - D'' \cdot \frac{(\sin^2\theta \cos 2\Omega)}{r^3} \tag{1}$$

where D′ and D″ can be related to ligand field parameters (or in the case of transition metal ions to g-values of EPR spectra), and r, θ and Ω are spherical coordinates of the proton (carbon or phosphorus) nucleus relative to the metal as origin, Fig. 1. In order to apply Eq. (1) it must be understood that it is derived on the assumption that (a) the magnetic moment is isotropic to first order (b) it is the second order terms which give rise to the anisotropy of susceptibility and that (3) there is a field inequivalence $x \neq y \neq z$. Note also that this rhombic field case is difficult to study unless the magnetic axes are known due to the large number of unknowns. The problem has not been solved effectively.

The above equation reduces to

$$\delta_p = D' \cdot \frac{(3\cos^2\theta - 1)}{r^3} \tag{2}$$

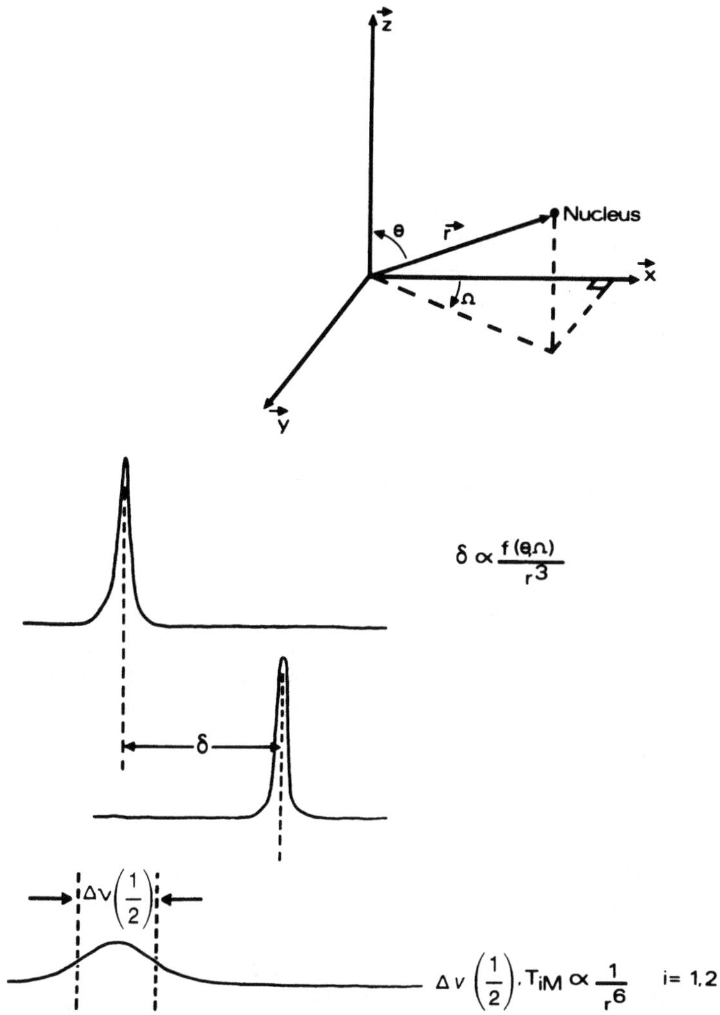

Fig. 1. Illustration of coordinates of a nucleus relative to Ln(III) and showing both shift *(middle)* and broadening effects *(bottom)* on NMR lines

in the case of axial symmetry, $x = y \neq z$. The symmetry here (and elsewhere) is not static symmetry of a given conformation but the time averaged symmetry over the ensemble[10–13]. Various rapid motions (on the NMR time scale) within a complex can reduce rhombic symmetry which would have been seen in individual conformations to this *effective* axial symmetry, see later. Thus the axial field equation may apply unexpectedly, which is helpful, but it does then refer to an average and interpretation must be cautions.

Equation (2) is still difficult to handle since it contains the quantity D' as well as the two numbers which we wish to determine, θ and r. If we select a resonance shift of a given nucleus and ratio all shifts of resonances of other nuclei in a given molecule to it, then in the ratios D' disappears. Thus for a rigid frame ligand there is the immediate possibility of carrying out a search for the exact geometry of the frame relative to the metal ion as origin. Apart from r and θ there are the unknowns of the major axis of symmetry of the magnetic atom, Ln(III). We require at least six shift ratios to begin to have confidence in our search calculation. A careful study of steroid binding to an organic shift reagent was done to explore the method using some dozen ratios[14]. The search was made employing computer graphics with full display of shift information. The derived structure was seen to be sensible in three ways: (a) it gave a sensible coordination geometry including the expected Ln(III) to coordinated oxygen atom distance; (b) it gave relative relaxation data for different protons using Gd(III) as a reagent in agreement with observation (see below); (c) the known rigid frame structure of the steriod was totally consistent with the NMR observations. Given care the procedure proved to be capable of generating conformations of such complexes of rigid frames with great accuracy[8, 14]. This validates the procedure using organic shift reagents and it was also shown that the symmetry axes were the same for all Ln(III) for the *organic* shift reagents.

II. Relaxation by Gadolinium

The gadolinium ion, Gd(III), has a half filled sub-shell and like Mn(II) and Fe(III) it has a long electron relaxation time and a spherically symmetrical ligand field. As a consequence of the slow electron relaxation rate it causes very considerable changes in the nuclear magnetic relaxation times, T_1 and T_2, of nearby nuclei but not shift. The latter change, T_2, is readily measured even by the broadening of NMR lines, Fig. 1. Since the Gd(III) is spherically symmetrical the relaxation enhancement due to the metal is given by

$$\frac{1}{T_{2M}} = \frac{1}{T_{2(obs)}} - \frac{1}{T_2} = f'(\tau_c)/r^6 \tag{3}$$

where $1/T_2$ is the relaxation rate in the absence of metal ion, r is the distance of the nucleus under study from the Gd(III) ion, and τ_c is the correlation time for dipolar interaction. $1/T_{1M}$ gives a parallel expression. Note that no contact contribution has been included. The equation permits direct determination of r if $f'(\tau_c)$ is known, as is frequently possible. Even without knowledge of τ_c ratios of $1/T_{1M}$ or $1/T_{2M}$ for different nuclei in a complex lead to a radial map of nuclei from Gd(III). This map can be used together with shift ratios in order to determine the structures of molecules. In the case of known outline structures it is also possible to use the relaxation data to assign resonances to particular nuclei. Mapping in this way proved extremely valuable in the study of lysozyme and led to the development of a range of relaxation probes which we and others[10–13] are constantly extending in biological studies, see Table 1.

Table 1. Relaxation Probes

(a)	Based on Gd(III): Gd(III), [Gd(III) · ATP], [Gd(III)(P_2O_7)], [Gd(III) · EDTA], [Gd(Acetylacetonate)$_3$]
(b)	Based on Cr(III): $[Cr(NH_3)_6]^{3+}$, $[Cr(CN)_6]^{3-}$, [Cr · ATP]
(c)	*Others* Mn(II), Fe(III), Cu(II), and their complexes; Magnetite particles.

The best relaxation probes are d^3, d^5 and f^7 configurations but all f^n ions can be used as isotropic probes and many d^n, especially d^1 and d^9, configurations can be used as anisotropic probes

These results prompt two immediate comments about the general suitability of the Ln(III) probe method. The first is that for molecules of relatively fixed frame, and this includes many proteins, where difficulty is experienced with more conventional NMR structural procedures, e.g. nuclear Overhauser effects (NOE) or coupling-constants, in deciphering the frame structure data, the use of Ln(III) reagents can be decisive. The second is that for mobile molecules *averaged* information can be obtained which can then be used in conjunction with other *averaged* NMR data to sort out the major contributing conformations and motions. It is extremely important to note that the different methods NOE, coupling-constant and lanthanide reagents are very different in kind so that when agreement is reached between them there can be little doubt as to the nature of the mobile system, see below.

III. Contact Shifts

It is usually the case that there is a considerable contact shift on the resonances of nuclei of atoms in the first coordination sphere of Ln(III) ions. The contact shifts can be separated from the dipolar shifts by quite simple procedures[15, 16] involving the study of complexes of the ligand with different Ln(III) assuming that the dipole shifts have axial symmetry, Fig. 2. The deduced contact shifts are quite useful in that they show which atoms are directly bound to Ln(III) in the complexes. We shall not concern ourselves further with contact terms in this article as no detailed structural information is available from them. The coordination sphere is usually generated by the searches using only dipolar shift and relaxation data.

IV. Some Theoretical Considerations

When the lanthanide shift measurements were first obtained the magnitude of the shifts, i.e. D′ in Eq. (2) for different lanthanides, were very puzzling to us. We turned to Prof. B. Bleaney for help and he provided a detailed theoretical explanation of lanthanide shifts based[7] upon second order contributions to the assymetry of the magnetic susceptibility, see above equations. The theory led to precise predictions of the relative sizes of the shifts to be expected from the different lanthanides[18]. The agreement between observation and theory in the test cases of a known frame (the lanthanide dipicolinate complexes for example, see Table 3) is quite remarkable and leads us to complete trust in Eq. (2) whenever such general agreement within the lanthanide series is obtained.

V. Curie Spin Relaxation

If we consider a paramagnetic ion, susceptibility, χ, with a very fast electron relaxation time τ_S such that $\tau_S < \tau_R$, i.e. not Gd(III), and where τ_R is the rotational tumbling time of the molecule, then relaxation times at a nucleus distance r from the metal have been shown to be given by the expression (4), which includes the applied field, B,

$$\frac{1}{T_{1,2}} = \text{Constant.}\ \frac{\chi^2 \cdot B^2}{r^6 T^2} \tag{4}$$

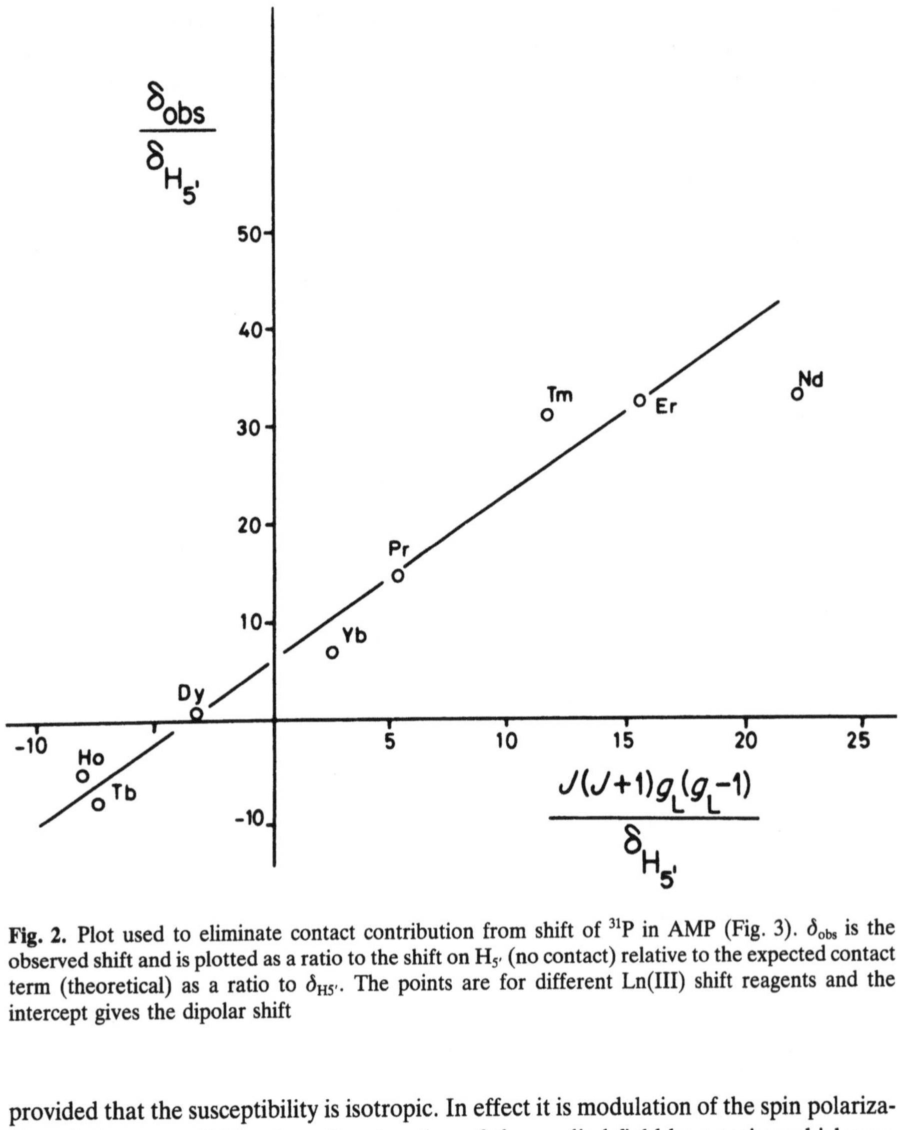

Fig. 2. Plot used to eliminate contact contribution from shift of ^{31}P in AMP (Fig. 3). δ_{obs} is the observed shift and is plotted as a ratio to the shift on $H_{5'}$ (no contact) relative to the expected contact term (theoretical) as a ratio to $\delta_{H5'}$. The points are for different Ln(III) shift reagents and the intercept gives the dipolar shift

provided that the susceptibility is isotropic. In effect it is modulation of the spin polarization of the susceptibility along the direction of the applied field by rotation which produces relaxation. For Ln(III) protein complexes, other than of Gd(III) this is the case since the condition $\tau_S < \tau_R$ holds for large molecules and thus relaxation is very susceptibility, field, and temperature dependent. The expression was first derived by Gueron[19] and by Vega and Fiat[20]. The theory was tested on haem compounds by Johnson et al.[21]. Lee and Sykes[22] have used the expression to study the calcium binding protein, parvalbumin, substituting Ca(II) by Ln(III). They chose to use Yb(III) since it gives relatively small contact shifts. It is very easy to find the line widths and their field dependencies. These expressions apply only to large molecules since for small molecules $\tau_S \simeq \tau_R$ and the conventional dipolar relaxation should be found for all Ln(III). The considerable advantage of this method is that it allows the constant in the equation for the dependence of relaxation on $(\text{distance})^{-6}$ to be determined empirically through the study of changes in relaxation with different lanthanides and with applied field. The ratio method still applies of course. The method is at its best for fully formed complexes in slow exchange – quite the opposite of the usual study using Gd(III).

C. Flexible Molecules

Let us turn next to an example of the use of Ln(III) in the study of more mobile structures, i.e. molecules which are flexible. We and others have stressed that any result we obtain using the above search procedures will be some average of real structures and that unless more than one method is used this average could be meaningless. The basic premise is that if several methods are used which employ different dependencies on r and θ etc. and they give the same *outline* conformation then it is correct and not an artefact. In our first work we chose to let the computer have freedoms in its exploration of conformation which knowledge of chemistry or other studies could have disallowed[23]. This was an ultimate test of the method. For example we did not fix bond angles at a saturated carbon as tetrahedral. We wanted to see what the structure search using shift and relaxation Ln(III) probes would produce. The molecule studied was 5′-adenosine monophosphate, Fig. 3. The flexible motions in the molecule were known both outside and inside the ribose ring. We allowed the computer to search over chemically reasonable and unreasonable structures both of the ring and the phosphate. (What is reasonable does not always turn out to be true!) We then published a picture of the outline "conformation" which this procedure generated, Fig. 4. It was not a single conformation but a family of related conformations about a centroid. Subsequently and in the light of extensive studies by Sundaralingham, Altona and many others[24], who did not employ the Ln(III) search methods, and which showed that there is present a particular biased ensemble of conformers we decided to carry out a second search using the lanthanide procedure. Now we restrained the bond angles to chemically reasonable values[25], an assumption made in the other studies[10–13], and we took into account coupling constant data, which, using the Karplus equation, gave the most probable combination of confor-

9-β-D Ribofuranosyladenine-5′-Monophosphate (AMP)

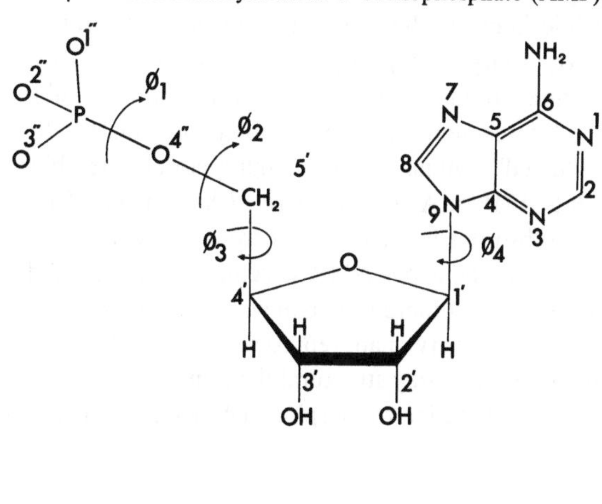

Fig. 3. 5′-AMP showing major flexible bonds but note that the ribose ring has preferred conformations

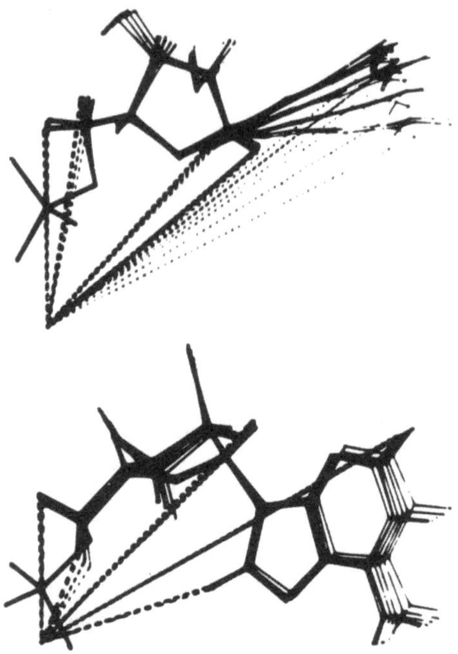

Fig. 4. Outline picture of 5′-AMP from a search of shift and broadening data only

mations for local regions of the molecule (two carbon) fragments of the structure. Further useful data were obtained from nuclear Overhauser effects. Agreement was now extremely good between all methods, and confined the molecule to certain heavily weighted conformations. We used around fifteen shift ratios and seven relaxation times for proton, carbon and phosphorus resonances. Additionally and very importantly our work showed that in solution the bases, which are used to make DNA and RNA, and the cyclic nucleotides do not change structure on binding Ln(III) and therefore we can presume that they do not change structure on binding Mg(II) or Ca(II), which are important in biology. Now the combination of the conformations which had to be used can be superimposed by a computer to give a blurred molecular outline which turned out to be much like the original "blurred" picture we gave from the direct Ln(III) search, Fig. 4. This being the case it is actually quite useful to have in mind the blurred picture while knowing the conformational mobility which lies behind it. (It is not possible to carry a large group of pictures in one's mind.) Conformations far from the "blurred" outline are in fact of high energy and not accessible.

Interestingly in the study of mono nucleotide phosphates bound in enzymes the "blurred" outline represents quite closely the "blurr" of electron density which represents AMP or ATP in the derived X-ray maps. The derivation of exact structural models for substrates in enzyme sites from X-ray diffraction data is often much less easy than obtaining the above NMR structures. Motion is difficult to portray on paper in the nmr structures but there is no real way to assess what averages of structure the X-ray derived electron density map represents either. Again very good agreement was achieved between the X-ray structural data on a protein phosphoglycerate kinase bound to magnesium adenosine triphosphate (ATP) in the active site, Fig. 5, and the same structural

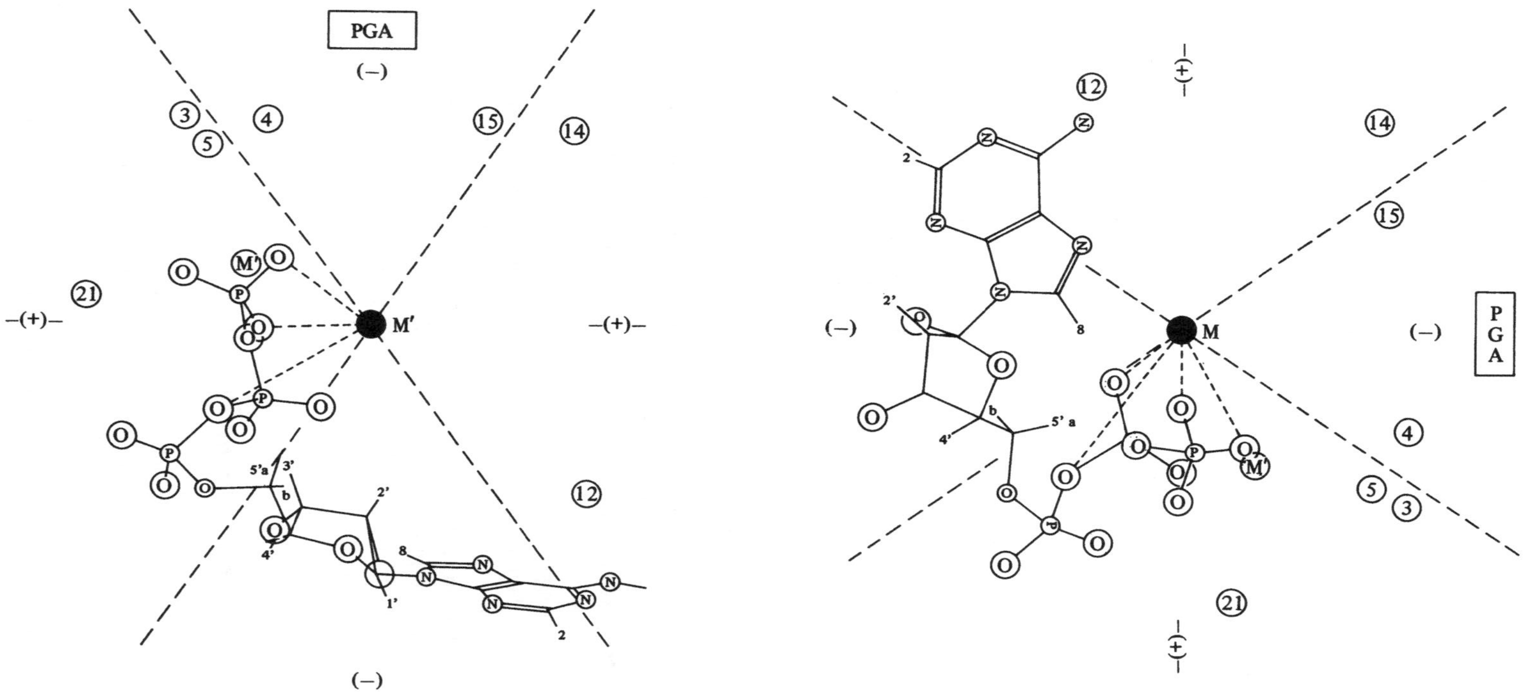

Fig. 5. The structure of the enzyme active site of phosphoglycerate kinase deduced without knowledge of the X-ray structure see Refs. 26 und 27 PGA is phospho-glycerate – the second substrate, numbers refer to individual resonances studied. ○ N; ○ water

data derived from NMR shift and relaxation data using Ln(III) complexes of ATP[26, 27]. No other study in solution can provide this type of information, which is essential if we are to have confidence in the use of crystal studies to give pictures of molecules in solution. We give other important examples later. We now summarise the way, as we see it, in which the lanthanide ions can be used as structure probes.

Given the uncertainties of solution studies, e.g. lack of knowledge of the exact coordination geometry and symmetry axes in complexes, progress can be made only by empirical procedures[10–13]. We have adopted the following rules as guide-lines to those circumstances in which effective axial symmetry (see below) holds. (N.B. The assumption of axial symmetry is critical to the method.)

(1) It is important to stress again that it must be demonstrated that there is a single site of binding of the Ln(III) ions.
(2) Contributions to the shift due to contact contributions involving the distribution of unpaired electrons on atoms close to the metal must be subtracted.
(3) Shift ratios must be the same within experimental error for several different Ln(III) ions, see below.
(4) Shift ratios must not be sensitive to temperature.
(5) The distances, *r*, must agree with distances which can be obtained from relaxation probe measurements.
(6) The magnitudes of shifts for different Ln(III) follow Bleaney's predictions.

If all the six conditions are met it is unlikely that the ensemble of states has contributions from widely different conformations. Given a set of shifts the average conformation of a molecule can then be found.

(7) It must be examined together with the conventional use of coupling constants and nuclear Overhauser measurements.

The use of the lanthanides in the study of the conformation of small molecules develops from their magnetic properties. It is often assumed that they bind without perturbing the structures. This assumption when not properly analysed has led to unwitting conflict between those who use the lanthanide method and those who do not since it is now clear that in some complexes, e.g. lipid headgroups, the lanthanide ions (and calcium) do change some of the conformational parameters of molecules. As stated above when the lanthanide method is used it is advantageous if other techniques such as nuclear Overhauser enhancements (NOE) and coupling constant analysis are used simultaneously. This can only be done properly if the NOE and coupling constant analysis are carried out on the molecule bound to a *diamagnetic* lanthanide ion. The lanthanide method is the more powerful the larger the molecule since the shift probe properties of the paramagnetic ion spread out the spectrum, giving new subspectra, at the same time as they give structural information.

I. Effective Axial Symmetry[10, 13]

All workers using paramagnetic dipolar shift reagents have recognised that the reduction of rhombic to axial symmetry is due to averaging by motions of ligands. A very good test of effective axial symmetry is that *all* Ln(III) cations give the same shift ratios since the three axes of the magnetic susceptibility in crystals are very differently related in size for different Ln(III) ions in the same environment, as can be readily shown especially by low-temperature measurements. The dipolar shift data which are then easily interpreted are those where the shift ratios for the different lanthanides are the same and the magnitude of the shifts follow expectation from the excellent theoretical framework provided by Bleaney. This is the case for *organic* shift reagents.

Difficulties arise where the shift ratios are not the same throughout the series. A frequently observed pattern is that the shift ratios for aqueous ions such as Ce(III), Pr(III), and Eu(III) are very similar but that ratios are very different for the elements Tm(III) and Er(III). The sizes of the shifts due to the ions of the first half of the series conform to expectations based on Bleaney's calculations but those of the second half do not. Two interpretations are possible. (A) The structure of the Ln(III) complexes, MY, changes from one part of the series to the next. (B) The structures of MY are the same but the dynamics are different. While Levine and Williams[28] considered (A) to be the most probable it has become increasingly clear from work by Dobson[29, 30] especially and then by Alsaadi, Rossotti and Williams[31] that (B) is correct although (A) may also be true. (B) says that rapid motional averaging out of rhombic terms is not effective for the smaller (heavier) aqueous lanthanide ions whether the structure is the same or not. The evidence is as follows, e.g. for Ln(III) · acetate complexes[28].

(i) Paramagnetic relaxation data (no angular dependence) give much the same distances from Ln(III) to different ligand atoms for all Ln(III) complexes of a given ligand.
(ii) At higher temperatures the shift ratios for all Ln(III) ions tend toward the values for the first members of the series i.e. Pr(III).
(iii) Rate processes are known to be slower for the smaller Ln(III) ions and shifts are generally larger. Averaging is expected to be more difficult. We return to this point below.

The question remains as to whether the hydrated complex ion structure is the same or different since although ligand coordination geometry is apparently roughly constant, see above, the hydration could well change. At the present time we do not know for sure if such complexes as the acetates $Ln^{3+} \cdot (^{-}O_2C \cdot CH_3)(H_2O)_n$all have the same value of n but there are strong indications that hydration changes, see below.

There are some curious observations as to when the Ln(III) cations all behave alike and when they behave differently as shift reagents. It would appear that such ligands as simple carboxylates and ester phosphates have shift ratios which change dramatically in the second half of the lanthanide series from approximately constant values in the first part of the series[28]. Clearly here, where the binding is weak and on/off rates are very fast $\gg 10^5$ per s, it is not the rate of ligand exchange in Eq. (iii) which is limiting averaging but the actual internal structure of the complex. We find that in some much stronger complexes e.g. $[Ln(dipicolinate)(H_2O)_6]^+$ where the ligand exchange is slow, averaging is fast enough to give an effective axial field i.e. averaging is the same on the surface of all the ions. Here even the resonances of free ligand are seen separately from the resonances of the complex[31]. The shift ratios for ligands such as EDTA are rhombic and very different for all Ln(III) but for adducts of Ln(EDTA) the shift ratios are axial in the first but not the second part of the series. Motion on the surface of Ln(III) in Ln(III) EDTA may be important. Finally if the ester-phosphates e.g. choline glycero-phosphate are incorporated in phospholipids in membranes[32] then the complexes (which are stronger than those of small molecule ester phosphates) are apparently axial throughout the series though the choline glycero-phosphate[33] complexes themselves in water are not axial throughout the series. *It is not the motions of the ligand or the ligand exchange rate which controls the appearance of effective axial symmetry.*

Taking all the evidence together it is clear that we need a deeper insight into the structures of the complexes $Ln(H_2O)_n$(ligand). So far we have used the paramagnetic properties of Ln(III) to look at Ln(ligand). We can of course examine the water bound in the complexes by similar methods. We shall take the series of complexes $Ln(dipicolinate)_n(H_2O)_m$ as examples below. Our results then throw light on the many problems of the nature of Ln(III) complexes in solution.

D. Water Proton Relaxation Rates[10–13]

Assuming that the relaxation of a given proton is due to the paramagnetism of the lanthanide ions then for the elements other than gadolinium the Solomon-Bloembergen equations for the water relaxation rates, $1/T_1$ or $1/T_2$, reduce to

$$\frac{1}{T_{1M}} = \frac{20}{15} \cdot \frac{\gamma_I \mu^2 \beta^2}{r^6} \cdot \tau_s \tag{5}$$

where τ_s is the electron spin relaxation time and r is the distance to the proton. The observed relaxation of water then depends on the mole fraction of water bound in the lanthanum complex, x_0/w, when the molar enhancement of relaxation (m.r.e.) is

$$\text{m.r.e.} = \frac{x_0}{w}\left\{\frac{1}{T_{1M}} - \frac{1}{T_w}\right\} \simeq \frac{20}{15} \cdot \frac{\gamma_I \mu^2 \beta^2}{r^6} \cdot \tau_s \cdot \frac{x_0}{w} \tag{6}$$

$1/T_w$ is the relaxation rate for water protons in the absence of the metal ion, μ is the paramagnetic moment of the ion and γ and β are conventional constants. We can now examine m.r.e. values in a series of complexes, in terms of μ, and calculating r from known radii of ions. This will reveal changes in τ_s or/and x_0/w. We commenced such a study using the Ln(III) complexes of dipicolinic acid.

E. The Structures of Ln(III) Complexes in Solution

I. The Dipicolinate Complexes[31–36]

It might be considered that the way in which to understand the water complex ion chemistry of the lanthanide ions in *aqueous* solution would be to start from their hydrates. In fact the hydrates prove to be most intransigent complexes. Their structures are still somewhat uncertain. I shall therefore start from a study of the tris-dipicolinates, i.e. tris 2,2′-carboxy-pyridine complexes, $Ln(dipic)_3$, about which a great deal is known.

Firstly the crystal structure of these complexes is known for very many of the different cations[37]. The complexes are nine coordinate throughout the series and there is no water in the first coordination sphere. As the radius of the central cation falls through the series the bond distance to the liganding N- and O-atoms changes but slightly. The first important points are that the high and constant coordination number of nine, the fixed geometry and the shorter bond-length to O- rather than N-atoms. (Differences from transition metal cations are obvious on each count.) Using proton NMR spectroscopy the structures can be examined in aqueous solution. The dipolar shifts on the *para* and *meta* protons and their ratios are given in Table 2, which shows that the shift ratios, $\frac{\Delta_{meta}}{\Delta_{para}}$ the structures are similar. The constancy of the ratio is consistent with axial symmetry for the complexes with the major axis in the expected direction. The absolute shifts in the series conform very closely to expected pattern, see Bleaney's values in column five, Table 3. The distances from the central metal to the two protons can be determined independently by the appropriate use of relaxation time studies and shown to be those

Table 2. Isotropic shifts of Ln(III) tris-dipicolinate complexes obtained from solution containing 1:2.5 and 1:4 mole ratio of metal:DPA measured at 270 MHz 300 °K

	1:2.5 mole ratio			1:4 mole ratio		
Metal ion	$\Delta^a p$ ppm	$\Delta^a m$ ppm	Ratio	$\Delta^a p$ ppm	$\Delta^a m$ ppm	Ratio
Ce	–	–	–	–	–	–
Pr	−4.01	−4.01	1.00	−4.01	−4.01	1.00
Nd	−2.21	−2.32	1.05	−2.21	2.32	1.05
Sm	−0.46	−0.46	1.00	−0.39	−0.39	1.00
Eu	+3.22	+4.13	1.28	+3.21	+4.24	1.32
Tb	−25.80	−29.59	1.15	−25.72	−29.58	1.15
Dy	−28.23	−32.71	1.16	−27.62	−31.43	1.14
Ho	−15.29	−17.87	1.17	+14.91	−16.98	1.14
Er	+8.74	+10.37	1.19	+8.61	−9.98	1.16
Tm	+17.97	+21.10	1.17	+17.54	+20.06	1.14
Yb	+6.59	+7.62	1.16	+6.47	+7.28	1.13

[a] Calculated with reference to the diamagnetic La(III)-complex

Table 3. Shifts (ppm) on para protons of dipicolinate

Metal	mono-dipic	bis-dipic	tris-dipic	Theory (Dy = 100)
Ce	−2.08			−6.3
Pr	−5.25	−5.16	−4.01	−11.0
Nd	−2.39	−2.51	−2.21	−4.2
Sm	−0.39	−0.34	−0.46	−0.7
Eu	+3.74	+3.57	+3.22	+4.0
Tb	−36.87	−35.12	−25.80	−86
Dy	−55.56	−58.91	−28.23	−100
Ho	−30.37	−22.79	−15.29	−39
Er	+17.05	+17.21	+8.74	+33
Tm	+33.70	+26.30	+17.97	+53
Yb	+14.53	+10.30	+6.59	+22

expected from the crystal structure within experimental error. Given these data the shift ratio can be understood. The structure is the same as in crystals.

Further use of relaxation data, now studying the water and the ligand protons[34–36], leads to an estimate of the outer sphere hydration of the lanthanides. We know there are no water molecules in the first coordination sphere of course. These outer sphere relaxation data for the different cations are proportional to susceptibilities and electron relaxation times and become very useful in the study of the inner sphere hydration of other complexes $M(dipic)_1(H_2O)_x$ and $M(dipic)_2(H_2O)_y$, see below. Note that there is no evidence of further association of the Ln(III) tris-dipicolinate complexes with small cations such as sodium ions. Later we shall show that these anions can bind to biological cationic surfaces and act as shift or relaxation probes.

We pass next to the Ln(III) bis-dipicolinate complexes, shown in Fig. 6. There is no crystal structure for the complexes. However detailed examination of the proton NMR spectra at room temperature shows that the shift ratios are again constant throughout the series and that absolute shifts follow Bleaney's predicted values, Table 3. The complexes must be isostructural and must have axial symmetry. Again the use of relaxation data gives an independent assessment of the relative distances of *meta* and *para* protons. We can put all the data together and give a structure for the complex ion as in Fig. 6 leaving three water molecules in the inner sphere. To prove that this is so we must analyse the proportions of the water, both bound and outer sphere.

Fig. 6. The structure of $[Lu(III)(dipicolinate)_2(H_2O)_3]^-$ in water from NMR data. Bond angles and distances are defined. *Hatched circles* are H_2O

After subtracting the outer sphere relaxation of the water, obtained from the data for the tris complexes, the excess relaxation of water protons is due to the inner hydration sphere. We can show then that three water molecules are part of the complex, using a comparison of ligand proton and water proton relaxation to eliminate τ_S and μ^2 in Eq. (6).

Dropping the temperature of the NMR study gives insight into the reasons for the apparent axial symmetry of a complex $[Ln(dipy)_2(H_2O)_3]^-$ which cannot have such a simple static symmetry. At very low temperatures the proton NMR spectra of the complexes show three separate one proton signals unlike the high temperature spectrum of a one-proton triplet, *para*, and a two proton double, *meta*, peak. Clearly at lower temperatures the two meta protons appear as individual peaks since there is not an effective axis of magnetic symmetry along the M–N bond. These signals can now be used to analyse the

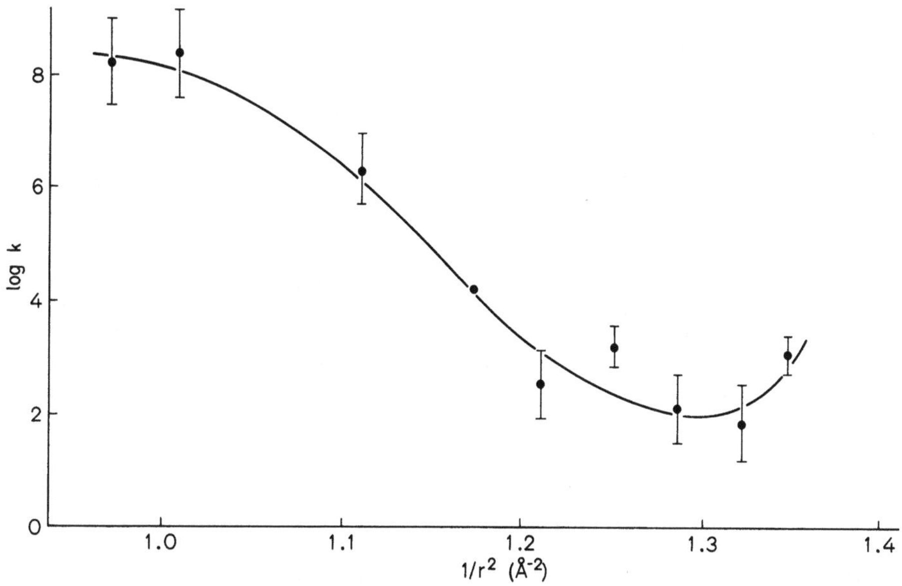

Fig. 7. Rates, k, of fluctional motion of dipicolinate in $[Lu(III)(dipicolinate)_2(H_2O)_3]^-$ at low temperature. Note the sudden change in log · k · with $1/r^2$

structure over a wide temperature range and to obtain the rate constants for rearrangement of the ligands on the surface of the lanthanide ions, using standard (NMR) equations for exchange. The rate constants, log k, for ligand fluctuations on the surface of the ions of radius r are given in Fig. 7.

We can turn finally to the mono dipicolinate complexes. The same analysis as above shows that in solution the M(dipic) complexes are isostructural. The exact structure has been determined using shift and relaxation data as above, see Refs. 34–36. Knowledge of the relaxation data for both ligand and water protons and the known relaxation of the contribution to water relaxation from the outer sphere then permits calculation of the number of water molecules in $M(dipic)(H_2O)_n$. We have shown that $n = 6$ for all the lanthanides.

Before continuing we look again at the relaxation data using different Ln(III)ions, leading to a description of the hydration of these and other complexes.

II. The General Problem of Hydration

We find that there is a good general correlation between μ^2 and $1/T_{1M}$, see Eq. (6), for those complexes in which we know that $x_{0/W}$ is constant, e.g. $[Ln(dipic)(H_2O)_6]^+$, and for outer sphere relaxation of water by $[Ln(dipic)_3]^{-3}$, see Table 4. However this relationship is not found for the simple aquo ions see below.

Now it is known that $Ln(III)(H_2O)_9$ can be formed by all lanthanide ions in crystals. Solution studies by X-ray diffraction indicate that the same tricapped trigonal prism exists in solution for Pr(III) and La(III) but there may well be variation in the relative

Table 4. Values of the relaxation enhancements of solvent water, MR1E, in presence of different Ln(III)-dipicolinate complexes measured at 20 MHz, 24 ± 1 °C

Metal ion	aquo s^{-1}	mono s^{-1}	bis-s^{-1}	tris s^{-1}
Ce	8.4 ± 0.2	4.8 ± 0.3	4.3 ± 0.4	2.3 ± 0.3
Pr	10.2 ± 0.2	6.6 ± 0.3	6.5 ± 0.3	2.6 ± 0.1
Nd	22.8 ± 0.5	18.5 ± 0.8	12.7 ± 0.5	3.2 ± 0.1
Sm	1.8 ± 0.0	1.4 ± 0.2	1.0 ± 0.3	0.1 ± 0.1
Eu	1.6 ± 0.0	1.3 ± 0.2	0.9 ± 0.3	0.1 ± 0.1
Tb	317.0 ± 6.3	241.0 ± 10.0	149.0 ± 6.0	58.0 ± 2.0
Dy	570.0 ± 11.4	378.0 ± 10.0	235.0 ± 7.0	108.0 ± 4.0
Ho	376.0 ± 7.3	268.0 ± 8.0	141.0 ± 5.0	37.4 ± 1.0
Er	398.0 ± 8.2	224.0 ± 5.0	110.0 ± 3.0	62.4 ± 2.0
Tm	414.0 ± 8.2	152.0 ± 5.0	56.7 ± 2.5	19.6 ± 1.0
Yb	53.1 ± 1.1	36.0 ± 2.0	15.7 ± 1.2	4.3 ± 0.2

Table 5. Relative shift on water protons for different Ln(III) aquo-cations (Er(III) = 1.0)

Ln(III)	δ ppm	Max. Dipolar (predicted)	J(J + 1)g(g − 1)
Pr	+0.09	−0.3	−0.21
Nd	+0.10	−0.13	−0.12
Sm	+0.005	−0.002	−0.32
Eu	−0.05	+0.12	–
Tb	−2.61	−2.6	+2.1
Dy	−1.71	−3.0	+1.85
Ho	−1.19	1.2	+1.48
Er	+1.0	+1.0	+1.0
Tm	+2.35	+1.6	+0.51
Yb	+0.66	+0.6	+0.17

Data from Levine B.A. D. Phil Thesis (Oxford) 1975

bond distances to the six water molecules of the prism and the three capping water molecules. However, the later Ln(III) ions are eight coordinate by the same methods, e.g. Tm(III)[38].

NMR studies of water proton shifts of Ln(III) solutions give the data in Table 5, see Ref. 39 as well. Correcting for contact terms as best as one can using the procedures of Ref. 15 again shows that there are anomalous pseudo-contact shifts around the Tm(III) ion, i.e. relative to expectation based on Bleaney's equations. The aquo ions are like acetate complexes but not like dipicolinate complexes. Here the symmetry, i.e. whether any of the complexes are effectively axial or not, cannot be decided with confidence but we observe that the deviations from expectation using an axial formulae for the dipolar shift are small for Pr, Nd, Ho, Dys – just where they are smaller for acetate complexes, and are large in the order Tm > Er > Yb again just as in acetate complexes. The structural anomaly is obviously present in both hydrates and monodentate complexes of ligands such as acetate. The conclusion is inevitable. The seires of hydrates, $Ln(H_2O)_x$, and series of monodentate complexes, $Ln(H_2O)_xY$ are not isostructural by NMR criteria i.e. they do not average to the same symmetry. We suspect that in some cases x changes but that in others the bond lengths and angles to water also change from the first to the second half of the series.

If it is asked why it is that the anomaly is greatest at Tm(III) then clearly it cannot be a matter of the radii or steric hindrance. There is only one additional factor which is the ligand field splitting. As shown in Fig. 8 the splittings are large at Tm(III) and greater than kT[38]. It is therefore likely that the structural changes are partly due to ion size changes and partly to ligand field effects. If we turn back now to the molar relaxation enhancement in water complexes then once again we observe anomalies in the Ln(III) series especially when we examine the first relative to the last few cations, Fig. 9. The difference does not arise from exchange rates but from changes in coordination geometry.

It would seem to be safe to draw the following conclusions. Where
(1) the values of shift ratios are very different in the second as opposed to the first half of the Ln(III) series;
(2) the value of τ_s for Tm(III) is long;

Then the hydrated complex has a different structure at the end of the series from that at the beginning. While at the beginning of the series axial symmetry arises readily, at the end of the series it does not.

If we accept these ideas we must assume that in the middle of the series there will be mixtures of complexes of different structures[38]. Note that it is not probable that the ligand geometries of flexible ligands in different Ln(III)Y will be precisely the same when the hydration of the Ln(III)Y complexes is different.

It is a matter of some concern that we do not know which ligands Y will generate effective axial symmetry in Ln(III)Y in the last part of the series while it would appear to be obtained almost invariably for Pr(III). Dobson, Delepine and Menear[29] in a careful study of the complexes of a dicarboxylate ligand which has a rigid frame showed that the Ln(III) ions of the second part of the series do not give axial complexes. They come to no firm, final, conclusions as to why this is. Now it is not just the physical structural properties which are of interest. Very similar problems have arisen in the discussion of stability constants and rates of reaction of Ln(III) ions.

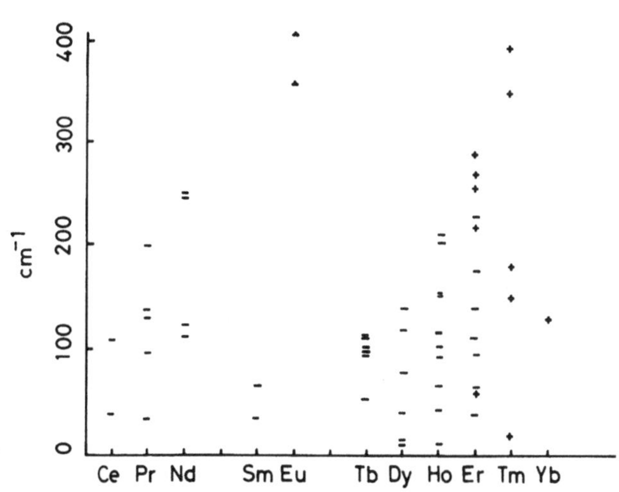

Fig. 8. Ligand field splittings in Ln(III) solids

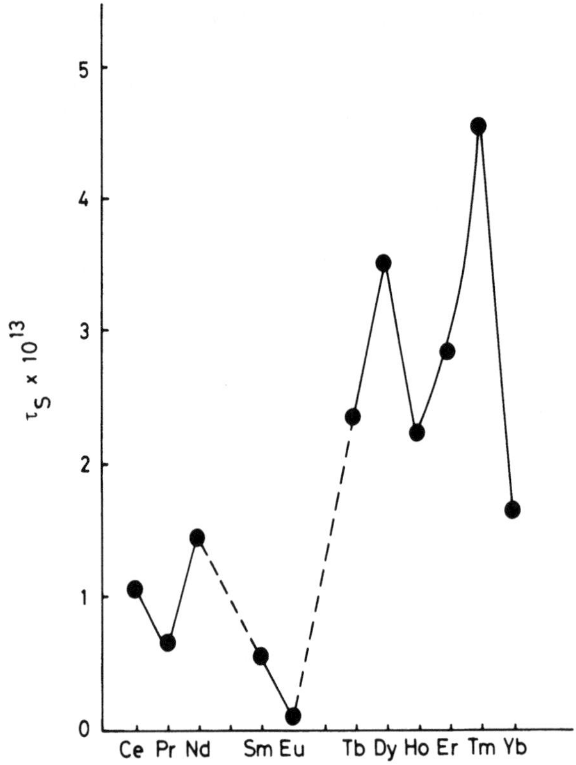

Fig. 9. Electron spin relaxation times ($\times\ 10^{13}$) for Ln(III) hydrates in water. Note the big difference between the first and second half of the series

F. Stability Constants

The stability of a complex at 25 °C has usually been discussed in terms of the ΔH°_{25} and ΔS°_{25}. We wish to discuss particularly these changes in Eq. (7) as the ion size of the cation changes.

$$-RT\ln K_{25} = \Delta G^\circ_{25} = \Delta H^\circ_{25} - T\Delta S \quad (7)$$

It is important to see first how steric constraints can affect the stability constants as the radii of cations decrease. Table 6 shows that in Group III A small ligands such as F^- or OR^- (OH^-, acetylacetonate, chromotropic acid) give complexes of increasing stability up to and including Al(III) and probably B(III)[40]. Large ligands such as carboxylates (EDTA and DTPA) show maxima in $\ln K_{25^\circ}$ between Lu(III) and Sc(III). Still more bulky anions, e.g. SO_4^{2-} and ester phosphates, give the most stable complexes with La(III). The changes are not seen in ΔS which almost invariably increases going to smaller ions despite the big differences in hydration number, $La(III)(H_2O)_9$ but $Al(H_2O)_6$. The ΔS terms must represent sums over both inner and outer sphere. It is the $-\Delta H$ term which usually shows the increasing effect of steric hindrance due to ligand size, Table 7, rising then falling with decreasing ion size in a series such as that of DTPA.

Table 6. Some group III A stability constants (log K)

	La	Gd	Y	Lu	Sc	Al
F^-	3.2	3.5	3.6	3.6	(6.0)	6.1
Acetyl acetone	5.1	5.9	6.4	6.4	8.0	8.6
Chromotropic acid		~6				17.4
EDTA	15.1	17.0	18.0	19.8	23.1	16.1
DTPA	19.0	22.2	–	22.1	–	18.5
(OH^-, S.P.)	21	23	24	25	30	33.5
Acetate	2.02	2.16	1.97	2.05		

Table 7. ΔH and TΔS (Kcal) for formation of DTPA complexes at 25 °C

	Al	Lu	Er	Gd	Nd	La
ΔH	+8	−5.1	−7.4	−7.8	−7.1	−5.2
TΔS	+34	+25.8	+23.7	+23.1	+22.5	+21.6

DTPA is diethylene-Triamine-NNN′N′N″-penta-acetic acid
TΔS does not reflect the inner sphere hydration Al(6), La(9)

A general account of these stability changes, ΔG, with ion size has been developed by Phillips and Williams[41]. It describes the observed patterns for idealised spherical anions of radius, r_A, and spherical cations, r_B, in the following terms. The free energy of hydration of anion plus cation follows the sum of $C_1/r_A + C_2/r_B$ where C_1 and C_2 are constants while the stability of the ion pair follows the function $C_3/(r_A + r_B)$. Simple analysis of complex formation in Group I A, II A and III A shows that for very small anions, e.g. F^-, OH^-, stability increases with decreasing r_B; for very large anions, e.g. ClO_4^-, it decreases with decreasing r_B; and for anions of intermediate size, e.g. RCO_2^-, it goes through a maximum. This is the behaviour we should expect for the lanthanide series. The reflection of ΔG changes in ΔH and TΔS is complicated however as we shall see. [Note the model describes water in terms of a bulk dielectric not a molecular solvent].

Direct determination of ΔH°_{25} and ΔS°_{25} by calorimetry shows that in many series of lanthanide complexes with anions, e.g. $EDTA^{4-}$ there is indeed a large entropy of stabilisation which increases with decrease of ion size. It also shows that a small favourable ΔH°_{25} often decreases somewhat with decrease in ion size. It is easy to see why these patterns of log K, ΔH and ΔS against atomic number are generally true of complexes in the lanthanide series but the details are more perplexing. Frequently the stability series show quite strong breaks at the half-filled shell ion and two weaker breaks roughly half-way between the half-filled shell ion and the beginning of the series and between the same ion and the end of the series – the so-called Tetrad-effects. These have been discussed in the following general (and simplified) elaboration of changes in ionic radii in the Ln(III) series, see Ref. 42. [Note this discussion is dependent on radius ratio considerations just as in the above. There is an alternative possibility, see below.]

The effect of filling the lanthanide f-shell on gas ion size is simply due to internal $4f$ electron-electron repulsion which is reflected in the L/S coupling, i.e. the magnetic moments of the ions, and in the repulsion between these $4f$ electrons and ligand electrons, a ligand field effect[42]. The angular terms of the ligand-field are probably of smaller consequence than the radial terms. As a consequence of these electron/electron repulsions the effective ion-size of the lanthanide ion decreases in a slightly uneven way and not just in two stages with a pause around Gd(III), the half-filled shell ion. Thus very minor changes in ion-size, changes in the incremental decrease of ion size with atomic number, cause the tetrad effect which correlates closely with magnetic quantum number changes although the magnetic energies are in themselves irrelevant to the energetics. Now given that the variation in ionic size is in four regular decremental waves we shall look more closely at how size is related to the thermodynamic constants, $\log \cdot K_p$, in the series. This involves an understanding of hydration on a *molecular* model both in the innner and outer sphere of the cations. The molecular model replaces the model used above in which water was treated as a bulk di-electric. In the latter case the properties of complexes can only change *regularly* with 1/r but this is not true of a molecular model.

I. Hydration Again

The chemist thinks of ion hydration in roughly the following way. Ions such as Al^{3+}, Ni^{2+}, Cr^{3+}, Co^{3+}, have defined inner sphere hydration numbers $[M(H_2O)_6]^{n+}$ and show relatively slow water exchange. Ions such as Na^+, K^+, Ca^{2+} have rapidly fluctuating coordination spheres with variable numbers of water molecules and very rapid exchange rates. This interpretation is no longer satisfactory since the neutron diffraction data of Enderby[43] show that the hydrates of sodium and calcium in strong solutions (see footnote 1 at end of the article, p. 118) are $[Na(H_2O)_6]^+$ and $[Ca(H_2O)_6]^{2+}$. However, our data on $[Ln(dipic)_2(H_2O)_3]$ shows that there is in fact rapid fluctuation on the surface of the Ln^{3+} (and surely on Na^+ and Ca^{2+}) ions. Putting the structures, the exchange rates and fluctuations together for all the hydrates of Groups I A to III A the picture of the hydration is one of very rapid motions of all kinds, both exchange on the surface and on/off the surface, though for say $> 75\%$ of all time the positions of oxygens relative to one another and to the cation are the same. The complexes have a dynamic structure. This motion in and around the ion will not be present for say $[Cr(OH_2)_6]^{3+}$ ions mentioned above. Now we expect the tightness of the inner sphere to have a profound effect also on the state of water molecules in the second and third (outer sphere) layers.

II. The Coordination States of Cations

If the inner sphere coordination is a highly dominant term in the lattice energy of salts of a metal ion then the crystal structures should be just the packing of complexes of a fixed M-dependent structure. For example metal hydrates of SO_4^{2-} give compounds $[M(H_2O)_6]^{n+}$ $SO_4^2 \cdot n \cdot H_2O$ for Mg^{2+}, Ni^{2+}, Al^{3+}, Cr^{3+}, Co^{3+}, Mn^{2+} etc. and similar hydrates appear in halides, and salts of other strong acid oxy-anions, e.g. nitrates. Even when water molecules do not occupy all sites around M the octahedron is maintained [Exceptions such as Co(II) and Zn(II) giving $M(O)_4$ polyhedra are easily explained on the basis of polarisation (ligand-field) terms but these are still central field effects, i.e.

Table 8. Structures of A-group metal salts

Some typical sodium salt structures

Salt	Na(I) Coordination No.	Na–O distances (Å) (min)	(max)
$Na_2CO_3 \cdot 10\ H_2O$	6	2.31	2.51
Na ascorbate	6	2.41	2.72
$Na_2PO_4 \cdot C_3H_7O_2 \cdot 5\ H_2O$	6	2.26	2.54
	6	2.28	2.59
$Na_4P_4O_{12} \cdot 4\ H_2O$	5	2.26	2.50
	6	2.41	2.53
$Na(CH_3(CH_2)_5COCO_2^-)$	6	2.37	2.67
$NaHCO_3$	6	2.38	2.48
$NaC_6H_7O_7$	6	2.30	2.46

Some typical potassium salt structures

Salt	K(I) Coordination No.	K–O distances (Å) (min)	(max)
K caprate	8	2.67	2.73
$K_2S_5O_{16}$	8	2.71	3.10
K picrate	8	2.74	2.92
$K_2[B_5O_8(OH) \cdot 2\ H_2O$	8	2.65	2.89
K polyphosphate	8	2.76	2.83
$K(HC_2O_4)(H_2C_2O_4) \cdot 2\ H_2O$	9	2.87	3.25
$K_2[C(COOH)COO]_2^-$	8	2.66	3.21
$KC_6H_4COOH \cdot COO^-$	6	2.62	2.98

Some typical magnesium salt structures

Salt	Mg(II) Coordination No.	Mg–O distances (Å)
Mg hexa-antipyrine · ClO_4	6	2.06 (Ca, 2.30)
$Mg(C_4H_8O)_4 \cdot Br_2$	6	2.16 (Oxygens, 4)
$Mg \cdot S_2O_3 \cdot 6\ H_2O$	6	2.05–2.12
$Mg \cdot SO_4 \cdot 4\ H_2O$	6	2.04–2.09
$Mg(HPO_3) \cdot 6\ H_2O$	6	2.01–2.12
$Mg_2P_2O_7$	6	2.00–2.11
$Mg(CH_3CO_2^-) \cdot 4\ H_2O$	6	2.00–2.10

Some typical calcium salt structures

Salt	Ca(II) Coordination No.	Ca–O distances (Å) (min)	(max)
$CaHPO_4 \cdot 2\ H_2O$	8	2.44	2.82
$Ca(H_2PO_4)_2 \cdot H_2O$	8	2.30	2.74
Ca 1,3-diphosphorylimidazole	6	2.26	2.36
	7	2.27	2.78
Ca dipicolinate · 3 H_2O	8	2.36	2.57
Ca $Na(H_2PO_2)_3$	6	2.31	2.33
Ca tartrate · 4 H_2O	8	2.39	2.54
$Ca(C_6H_9O_7)_2 \cdot 2\ H_2O$	8	2.39	2.47

centred on M^{n+}]. The bond lengths from M to 0 or M to N are always fixed ± 0.10 Å, see Table 8 for Mg^{2+}. The contrast with the complexes of Na, K, Ca, Ln could not be greater. Table 8 shows that their coordination spheres are not maintained in different salt crystals and there are no regular expected bond distances. *This means that central field effects do not dominate bond lengths and bond angles.* In other words the outer sphere energies are commensurate with and cooperative with the inner sphere to a much larger degree for larger A-subgroup cations. There is no fixed inner or outer sphere pattern. Even the distinction is lost. It is then exceedingly probable that the variations of stability constant in Ln(III) series is not just a problem of changes in the inner sphere but depends upon the outer sphere interactions with M and the inner sphere ligands. The idea of a *radius* disappears.

The outer sphere changes will undoubtedly effect not only changes in ΔH and ΔS but in ΔC_P as ΔC_p is large for water itself. If the above description of the nature of lanthanide ions is correct then a structural explanation of their properties must be replaced by a statistical mechanical picture. It is this picture which I wish to carry over into the discussion of the uses of lanthanide ions as probes in biological systems, and therefore I wish to examine it in more detail.

III. Step-Stability Constants of the Dipicolinates[44)]

Table 9 shows the trends in the step stability constants for the dipicolinates. Firstly we observe the rise and fall in stability constants from La to Al in all three steps. Clearly as expected steric hindrance is an increasing problem and increases from mono to bis to tris complexes. Now a second way in which to analyse steric hindrance is to look at log K_n – log K_{n+1}. Curiously we find that log K_1/K_2 for La is 2.2 but for the smaller ions Lu and even Al it is 1.2. For log K_2/K_3 the values are La, 1.5, Dy 1.6 and Lu 3.1 as expected. Smaller step-constant differences for smaller ions cannot be explained on radius ratio considerations (but in coordination changes) and demand a look at ΔH and ΔS in molecular terms. In fact closer inspection of log K_1 itself shows that it rises rapidly from La to Pr but it does not rise again until after Dy, log K_2 rises more steadily through the series. Generally for multidentate ligands log K_1 behaves most anomolously in these Ln(III) series.

Table 9. Stability constants (log K) for dipicolinic acid complexes

	La	Pr	Gd	Dy	Y	Lu	Al
log K_1	8.0	8.6	8.7	8.7	8.5	9.0	4.9
log K_2	5.8	6.5	7.3	7.5	7.3	7.8	3.4
log K_3	4.3	4.8	5.8	5.9	5.6	4.7	0

IV. Thermodynamic Data for Ln(III) Dipicolinate Complexes

Now that we have established the formulae and structures of the complexes $Ln(dipic)_3$ and $Ln(dipic)_2(H_2O)_3$ throughout the lanthanide series and have looked at general factors affecting stability constants we can re-examine the stability constant and thermodynamic data for the following reaction (8) in detail and in structural terms remembering that outer sphere hydration is not yet written into this equation.

$$Ln(dipic)_2(H_2O)_3 + dipic \leftrightarrows Ln(dipic)_3 + 3\,H_2O \qquad (8)$$

Table 10 gives the data for $\Delta G^\circ_{25^\circ}$ (RTlnK), $\Delta H^\circ_{25^\circ}$ and $T\Delta S^\circ_{25^\circ}$[44]. The fall in $-\Delta G_3$ toward the end of the series after the initial rise clearly reflects an increasing steric hindrance in the coordination sphere. Quite unexpectedly however there is a very large change, a fall, in ΔS_3, after the ion Tb(III). Seeking an explanation for these changes it is clear that the sudden loss of ΔS_3 of stabilisation after Gd(III) is in fact the major cause of loss of stability since the change in $-\Delta H_3$ is favourable to complex formation and considerable in just this part of the series. Returning to the NMR study we observed that internal motion was increasingly restricted *in* the complexes of the smaller ions. It is not the change in the number of coordinated water molecules or of free ligand molecules which is dominant in ΔS_3 since this term is constant. The critical factor would seem to be the loss of internal motion of the ligands in $Ln(dipicolinate)_3$ after Gd(III) and probably in loss of motion of outer sphere hydration. (We must observe that the reaction here is between ions of like charge and not unlike charge but as the complex ions are identical and very large we would not expect anything approaching the above unfavourable changes in ΔS_3 in a series).

In the light of these observations we can go forward to an analysis of the reaction

$$Ln(dipic)(H_2O)_6 + dipic \leftrightarrows Ln(dipic)_2(H_2O)_3 + 3\,H_2O \qquad (9)$$

where ΔS should be increasingly favourable along the series since this is a reaction between ions of opposite charge. Turning directly to the thermodynamic data we find in Table 10 a very different picture from that for the formation of the 2:1 from that for the 3:1 complex.

In Table 10(b) $-\Delta G_2$ rises to the end of the series, $-\Delta H_2$ falls from Tb(III) as expected if there were steric constraints, while it is now $T\Delta S_2$ which rises rapidly in the central and last part of the series. Here we know that although steric hindrance is rising in the second half of the series both in the complexes $[Ln(dipic)]^+$ and $[Ln(dipic)_2]^-$ there is completely free motion at room temperature. The simplest explanation of the rise in ΔS_2 is that in these complexes, as opposed to those at the end of the $[Ln(dipic)_3]^{3-}$ series of complexes, the way in which $T\Delta S$ and ΔH compensate, within the complexes and as seen in but rather small changes of ΔG, has changed. We can explain this as follows.

Examination of Tables 10(a) and 10(b) show clearly that ΔG is more constant than ΔH and $T\Delta S$, which move in opposed senses and in either direction. This is a common observation in the thermodynamics of ion-pair formation. The so-called compensation of $T\Delta S$ and ΔH arises since when the metal and the ligand come together there are generated two possible changes (i) the M–L bond is made as short as possible, $-\Delta H$ is large, and motion is restricted by steric hindrance, $+T\Delta S$ is small, or (ii) internal motion is kept as high as possible, $+T\Delta S$ is large, but of necessity the bond-length is now larger, $-\Delta H$ is small. The same value of ΔG can thus arise in two ways. No changes in coordination numbers are as yet involved. Observed ΔH_T and ΔS_T are now temperature dependent if we use the equation $\Delta G_T = \Delta H_T - T\Delta S_T$ to determine them. In fact at

Table 10. Stepwise thermodynamic data for the formation of Ln(III) complexes of dipicolinate

(a) Tris-Dipicolinate Complexes

Ln(III)	ΔG(kals/mol)	ΔH(kals/mol)	ΔS(e.u.)
La	5.79	2.23	12.0
Ce	5.93	3.15	9.3
Pr	6.54	3.51	10.1
Nd	6.84	3.77	10.3
–	–	–	–
Sm	7.25	4.29	9.9
Eu	7.44	4.61	9.5
Gd	7.89	4.83	9.8
Tb	7.98	5.80	7.3
Dy	8.01	6.64	4.5
Ho	8.08	7.39	2.6
Er	7.70	7.65	0.1
Tm	7.38	7.40	0.0
Yb	6.87	7.12	−0.9
Lu	6.28	6.66	−1.2
Y	7.53	6.92	+2.0

(b) Bis-Dipicolinate Complexes

Ln(III)	ΔG(kals/mol)	ΔH(kals/mol)	ΔS(e.u.)
La	7.88	3.18	15.5
Ce	8.23	3.60	15.5
Pr	8.76	3.94	16.2
Nd	9.09	4.10	16.7
–	–	–	–
Sm	9.50	4.69	16.2
Eu	9.65	5.05	15.4
Gd	9.90	5.15	15.9
Tb	10.05	5.32	16.4
Dy	10.16	5.00	17.3
Ho	(9.96)	4.22	19.2
Er	10.33	3.90	21.6
Tm	10.45	3.62	22.8
Yb	10.52	3.86	22.3
Lu	10.54	3.80	22.5
Y	9.85	3.88	20.1

(c) Mono-Dipicolinate Complexes

Ln(III)	ΔG(kals/mol)	ΔH(kals/mol)	ΔS(e.u.)	
La	10.83	3.12	28.8	24.2
Ce	11.32	3.55	26.1	24.8
Pr	11.70	3.91	26.1	
Nd	11.91	4.01	26.5	25.7
–	–	–	–	–
Sm	12.01	4.28	25.9	24.7
Eu	11.99	4.07	26.6	
Gd	11.87	3.58	27.8	25.3
Tb	11.80	2.69	30.5	
Dy	11.83	2.17	32.4	28.7
Ho	11.85	1.95	33.2	30.2
Er	11.94	1.85	33.8	31.8
Tm	12.03	1.83	34.2	
Yb	12.05	1.92	34.0	33.7
Lu	12.29	2.19	33.9	
Y	11.52	1.44	33.8	

different temperatures motional or bond length considerations can dominate in quite different ways for one and the same metal ion. This is normally proven by examining the temperature dependence of ΔH found in ΔC_ϱ ($\Delta H_T = \Delta H_0 + T\Delta C_\varrho$). As a consequence ΔG_T may reflect ΔH_0 (0 °K) more than ΔH_T does. Note again that ionic radii will not vary smoothly with atomic number.

The very precipitate changes in ΔS_T and ΔH_T which occur over very small changes in apparent radius arises from the cooperative interaction within the inner and outer coordination spheres involving large numbers of molecules. This means that electrostatic interactions have considerable selectivity in overall structure hidden behind very small differences in ΔG. Since in biology structure and motion within structure are as important as stability we shall need to keep this feature in mind.

Returning to the dipicolinates. We conclude that stability is made as high as possible in $[Ln(dipic)_3]^{3-}$ by making bond lengths as short as possible at the end of the series but keeping mobility as high as possible in the first part of the series. In the $[Ln(dipic)_2]^-$ series stability is made as high as possible by keeping as much motion as possible throughout the series. We shall find these effects differentially in many Ln(III) complexes. It is then ΔG_T and not ΔH_T that must be correlated with parameters of the Ln(III) ions, e.g. radius and electronic structure, in order to understand overall stability but this will give a wrong structure picture if we are not careful.

In the formation of the series of complexes $[Ln(dipic)]^+$ from the aquo ions the situation is still more complicated since we do not know with certainty the hydration of $Ln(H_2O)_n$ at the end of the series, though it is probably n = 8, while at the beginning n = 9. Table 10(c) shows now that $-\Delta G$ falls in the middle but rises at the end of the series, while $-\Delta H$ goes through a maximum and a minimum and $T\Delta S$ falls then rises. We can only suggest an explanation. In the hydrate $[Ln(H_2O)_n]^{3+}$ motion is restricted but bond lengths shorten toward the end of the series as *n* goes from 9 to 8. Thus from Eu to Tm $-\Delta H$ of hydration increasingly reduces $-\Delta H_1$ of reaction. At the same time motion in the hydrate is increasingly restricted and this increases $T\Delta S$ of reaction. Motion is roughly the same in all the complexes $[Ln(dipic)]^+$. In this way we can account for the thermodynamic data including the peculiarities of the step constant differences *and* the NMR structural data which indicate gross structural and kinetic similarities in $[Ln(dipic)]^+$ but very different behaviour in $[Ln(H_2O)_n]^{3+}$ toward the end of the series. We shall see that this explanation is consistent with the behaviour of monodentate ligand complexes, where we believe that there is a structural change in the second half of the series both in $Ln(H_2O)_n$ and $Ln(H_2O)_nY$ but it is not necessarily identical for both.

In Table 11, ΔG (mono-acetate) fluctuates through the series, rising and then falling while $-\Delta H$ again rises gradually to Sm(III), favouring complex formation and then falls precipitously to Dy(III) with a corresponding compensation in ΔS. Even in the formation of the bis-acetate from the mono-acetate the same precipitous change is found now between Dy and Yb. There are also structural breaks following the middle of the Ln(III) series which affect many properties and are strikingly shown in NMR shift ratios. Discussion of tetrad or similar effects is likely to be grossly affected by changes in bond distances, coordination numbers, and ligand mobilities.

To summarise, and in general agreement with the discussion of Grenthe[45)] and Phillips and Williams[41)], in a series of cations of changing „radii" the interaction free energy of the cations with a given anion can not be understood without reference to the states structure and mobility, of the water and the ligand in both the free ion and the complex.

Table 11. Thermodynamic data for the Ln(III) mono-acetate complexes

Ln(III)	$-\Delta G$(kcal/mol)	$-\Delta H$(kcal/mol)	ΔS(e.u.)
La	2.16	−2.18	14.6
Ce	2.33	−2.09	14.8
Pr	2.50	−1.72	14.2
Nd	2.63	−1.71	14.6
–	–	–	–
Sm	2.77	−1.45	14.2
Gd	2.54	−1.86	14.8
Dy	2.33	−2.92	17.6
Ho	2.28	−3.17	18.3
Er	2.24	−3.27	18.5
–	–	–	–
Yb	2.30	−3.51	19.5

Because of steric factors there is not a simple relationship between ΔG or ΔH or $T\Delta S$ and the ionic "radius". There is then no simple explanation of such observations as tetrad effects, see below, since some such breaks could come about in an apparent series of ion size changes of *equal* increment. Tetrad breaks may occur at the expected place following „ionic size" differentials but may also occur elsewhere through the sensitivity of packing problems.

We can confirm this point of view by looking at paramagnetic shifts and τ_s (for water) in the complexes of supposedly different water coordination. We find that in the Ln(III) series of complexes, LnY, anomalies often occur at around Tm(III) especially. We conclude that the exact structures of Ln(III) complexes like those of Na(I), K(I) and Ca(II) can not be represented by single simple pictures and we must refer to statistical populations of complexes of different structure at equilibrium. The complexity of the structural and dynamic features of this A-subgroup chemistry has been used to functional advantage both by man in cements and plasters and by biology in shells and bones and in messengers.

G. Kinetics of Ln(III) Reactions[45, 46]

The understanding of rates of reactions of cations in solution was much advanced by the fast relaxation techniques of Eigen[46] and the subsequent discussion of substitution often in terms of the rates of loss of water from the hydration sphere. It is generally possible to write two pathways of complex ion formation

Dissociative $\quad M(H_2O) \rightarrow H_2O + M \xrightarrow{L} ML$

and

Associative $\quad M(H_2O) + L \rightarrow \underset{\text{outer sphere complex}}{M(H_2O)L} \rightarrow ML + H_2O$

Relaxation of complicated ligands may occur as a step in both pathways. Diebler and Eigen [46)] indicated the ways in which such mechanisms could be analysed using fast reaction methods. Several studies of Ln(III) complex formation and of the formation of Ln(III) mixed complexes have been analysed. Generally the dissociative mechanism is considered to dominate and we are then concerned with the water exchange rate. Several studies have shown that the rate decreases from La(III) to Lu(III) but there seems to be a minimum rate around Tm(III). This is also seen in the rate of rotation of ligands on the surface of the ions, Fig. 7. There may be a small crystal field term, or another contribution to a „tetrad"-like effect from the 4f electron core. However in the hydrate the precise relationship between the inner and outer sphere water may also be important as we saw when we discussed the heat and entropy of complex ion formation.

Examination of the structures of Ln(III) hydrates in crystals and our knowledge of Ln(III) complexes in solution now throws up a problem which the above equations do not readily meet. There is no certain distinction between inner and outer sphere for ions such as Ln(III). Firstly the inner sphere is constantly switching between 8- and 9-coordination but 9-coordination is not far from 6-innermost water molecules which can distort to an octahedron and 3-outermost water molecules. The steps of kinetics can involve multiple re-arrangements of the cation hydration shell which is itself variable in the series of Ln(III). The model equations above are only guides to thinking.

This is important in biological systems since neutron diffraction data indicate that the structures of Na^+ and Ca^{2+} are $[M(H_2O)_6]$ octahedron in strong solutions (see end). Yet many compounds of both these ions have higher coordination numbers.

I. Water Molecules Around Metal Ions

This essay on the lanthanides has repeatedly drawn attention to a problem of immense importance in both chemistry and biochemistry. The role of water in controlling the stability, the structure, and the lability of coordination compounds. In fact the emphasis extends from coordination compounds to the surfaces of solids[47)]. The role of water is then bound to be extremely important not only in complex chemistry and catalysis but in the growth and properties of crystals and amorphous materials. We can illustrate the problems outside Ln(III) chemistry[48, 49)].

It is known that a vast variety of enzymes use metal ions in acid/base catalysts. In some cases the role of the metal is to activate water directly, e.g. $Zn(OH)_2$ becomes $Zn(OH^-)$ in carbonic anhydrase, but in others it may be that the metal just forms a particularly constructive (useful) H-bond network, e.g. calcium in phospholipase A_2 and in *staph.* nuclease. Substitution of one metal by other metals is now a critical test of the precision of the catalytic site and we know that nickel does not substitute for zinc in carbonic anhydrase, although it binds, and that Sr(II) has a different activity in lipases and nucleases from Ca(II). It is the water in the coordination sphere which is partly responsible for these changes.

Again water is found to be present in bone as it changes from largely $Ca(H_2O)_n(HPO_4)$ to $Ca_2(OH)(PO_4)$. The H-bond network is critical. Mg or Sr substitution affects bone in different ways and, as yet, we understand little of these changes. It is the hope of the author that the study of Ln(III) complexes will show the factors involved in the role of water around cations extending from the first to the second coordination sphere. Maybe at more acid pH values than biology can use, man can make very refined acid catalysts from the Ln(III) ions for use in water, and can vary selective action by changing radius in very small steps – a possibility not open to biology except by using strain in large polymers, proteins.

In the A-subgroups of the Periodic Table it appears that the first coordination sphere dominates the energy of the aqueous complexes for ions of about 0.9 Å or less e.g. Li^+, Mg^{2+}, Sc^{3+} but not of larger ions such as (Na^+) K^+, Ca^{2+}, La^{3+}. This means that there are rapid changes in crystal or solution structures both in solid and solution states on going to the larger elements. It may well be that these changes are also responsible for large differences in the rates of rearrangement around the large as contrasted with the small cations. The faster rates of rearrangement for Na^+, Ca^{2+} and La^{3+} (not simply related to ion size but to a lattice arrangement stemming from ion size) could be of extreme value in the fashioning of materials such as bone, shell, plaster and concrete in all of which calcium is the major cation, as well as in cooperative mechanical trigger devices in biology such as membrane fusions, muscle contractions and so on. Here again calcium is the cation of choice but Ln(III) was not readily available. Note that changes in structure, thermodynamic properties and rate constants are not regular functions of radii but as in Fig. 7 can undergo quite rapid, almost step, changes. The explanation of the so-called Tetrad effect is complicated by the nature of intermolecular interactions in complexes.

H. Uses of Lanthanide Ions in Biological Studies

The lanthanide ions can be used to examine many of the potential properties of the A-sub group ions of the Periodic Table. The following is a list of some uses and properties of lanthanides which can be readily analysed

(i) Heavy atom "stains" in electron microscopy or X-ray diffraction studies
(ii) Thermodynamic properties using lanthanides as competing ions for specific sites, e.g. calcium sites.
(iii) Kinetic properties – the rates of exchange of ligands on and off ions and ligand mobility on ion surfaces (fluctional properties); rate processes internal to ligands
(iv) Structural features using NMR, ESR, or energy transfer leading to fluorescence[50–52].

I. Lanthanide Ions in X-ray Diffraction and as Electron Microscopy Stains

The simplest way in which to use a lanthanide ion is as an electron dense material. The best known use in the study of crystals is that of Ln(III) as a scattering element replacement for Ca(II) or Mg(II). Examples are the structures of concanavallin A and t-RNA. The replacement is not always strictly isomorphous and care is required in interpretation. Some proteins bind Ln(III) although they function without metal ions, e.g. lysozyme, and use has been made of the cations in X-ray studies of this enzyme. The early comparisons of lysozyme structures in solid and solution states were based on the use of Gd(III) as a probe in the enzyme for NMR studies, see below.

Lanthanides have been used for a long time in electron microscope (EM) studies as they stain anionic surfaces e.g. membranes. Again parallel NMR studies have shown that the membranes are perturbed slightly (see below) but not significantly at the EM-level. An alternative mode of staining is to use $Ln(FOD)_3$ where Ln is any lanthanide and FOD is bis(5,1-trifluoro) acetyl acetone, the chelating agent used in shift reagents. This chelate compound stains the membrane internally as it is hydrophobic. An example of its use in staining vesicles is shown in Fig. 10. Once again NMR studies have shown that the reagent does not perturb the membrane grossly[53)].

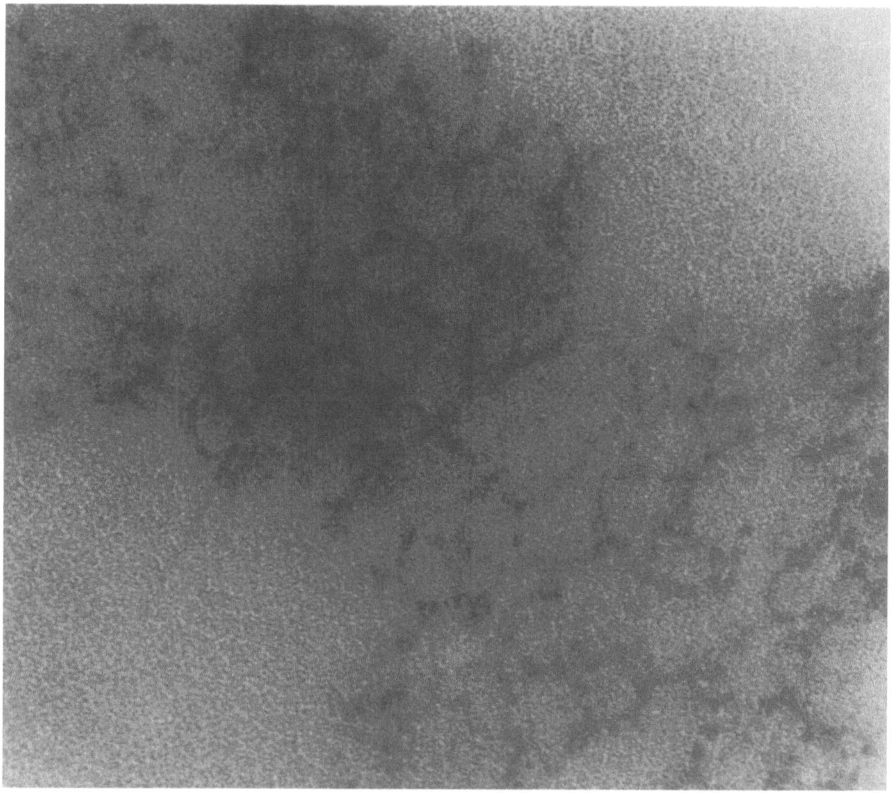

Fig. 10. Electron microscope picture of vesicles (500 Å radius) stained with $Gd(FOD)_3$; Magnification 5×10^5

II. New Ln(III) Reagents

The major reagents used in the study of both small and large molecules have been either Ln(III) aquo cations or $Ln(III)(FOD)_3$ complexes[54)]. It has been found that Ln(III)(EDTA) complexes are valuable reagents for anionic centres and that

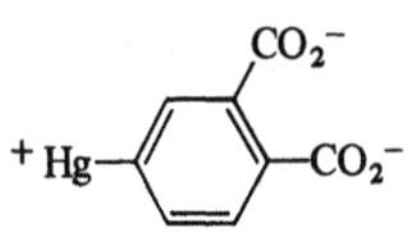

$Ln(III)(Dipicolinate)_3^{3-}$ are useful for cationic surfaces but greater specificity and selectivity of binding of Ln(III) probes is required[55]. A promising approach[56] is to use the above reagent which binds to sulphydryl groups so locating the dicarboxylate Ln(III) binding group. An alternative is to use the reagent synthesised by Desreux[57], formulae I, Fig. 11. Many years ago it was suggested that a variety of ring chelates could be used to bind metal ions e.g. formulae II[58]. However not until the accidental development of III[59] and then the preparation of cryptates by Lehn[60] did this chemistry became realised. The Desreux compound which we have now used extensively is particularly advantageous in that (1) it is water soluble (2) Ln(III) does not dissociate readily from it and (3) it binds to hydrophobic parts of molecules e.g. to the indole ring of tryptophan and not to the $-NH_3^+$ or the $-CO_2^-$ groups. It is known to be a very good probe of protein surfaces. It can be used to stain or as an NMR reagent.

What is required now is to add functional groups to Desreux reagents so that they can be linked to specific substrates. In this way they will probe specific sites instead of general hydrophobic areas. There is little doubt that the use of Ln(III) reagents can be increased greatly through organic synthesis of special Ln(III) ligands.

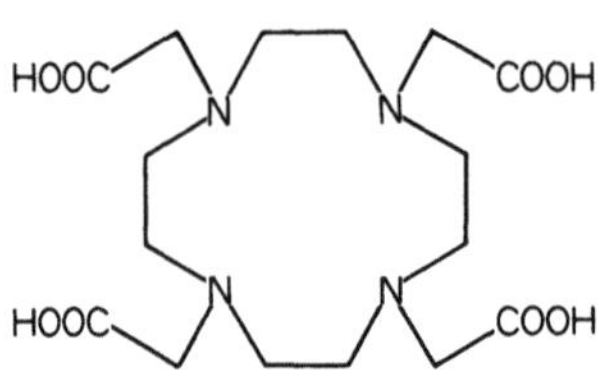

(I)

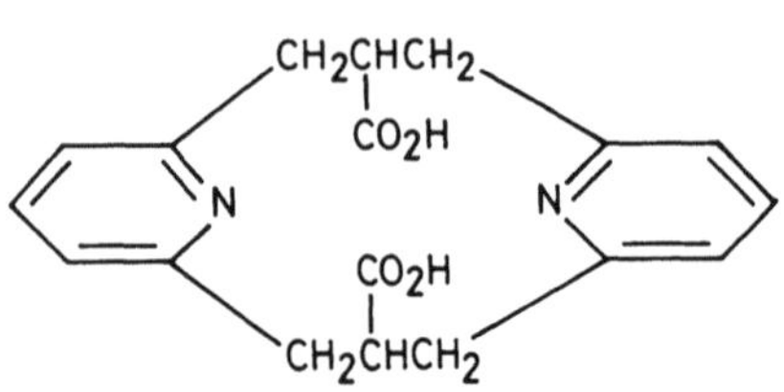

(II)

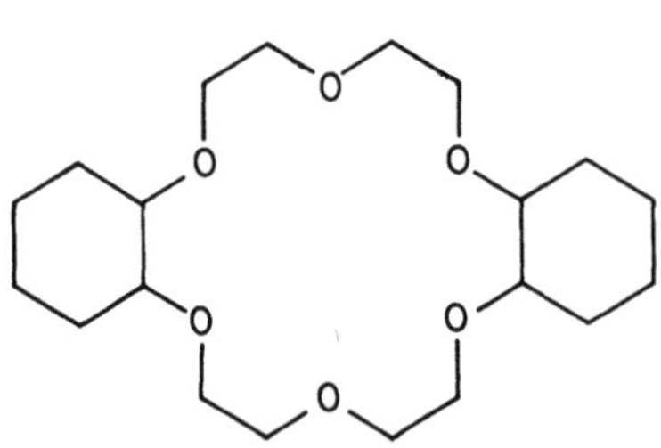

(III)

Fig. 11. The formulae of possible chelating agents for the development of Ln(III) probes as specific reagents

III. The Use of Ln(III) Ions as NMR Probes of Biological Structure[54, 61, 68]

I have already given an outline of the procedures and turn immediately to applications.

a. Lipid Bilayers[62]

The lipid bilayer is most readily studied in vesicles of about 500 Å diameter. The bilayer can be made of many lipids but the most common is lecithin, phosphatidyl choline (PC), which is shown in Fig. 12. Packing the lipids in the vesicle puts some 66% of the lipid headgroups on the external face and some 33% on the internal face. The head-group of this lipid, PC, is phosphocholine $-O-PO_2^{\ominus}-O-CH_2 \cdot CH_2\overset{\oplus}{N}(CH_3)_3$. Thus the head-group can be studied by P, H, D or C NMR. The long fatty chains can be looked at by C, H, or D NMR as can the glycerol adjoining link. The studies of the conformation of the molecules have been made using conventional spin-decoupling (and the Karplus equation), nuclear Overhauser effects, and by Ln(III) probes. Direct comparison with crystal structure data shows close agreement between solution and X-ray data but it is only possible in the absence of the Ln(III) cations since it has become clear that the Ln(III) cations alter the conformation of the head-group, see below. It is extremely likely that all bound cations also alter the head-group geometry in a similar way. In the unbound form the head-group lies roughly in the plane of the membrane as might be expected from electrostatic interaction of $-N(CH_3)_3^+$ and $-PO_2^-$– units. The metal ions bind to the $-PO_2^-$– groups and are presumably repel the $-N(CH_3)_3^+$ away from the lipids, Fig. 12. Details of these structural studies are given by Hauser [63] et al., and Seelig and coworkers[64]. They were determined by shift and relaxation probes.

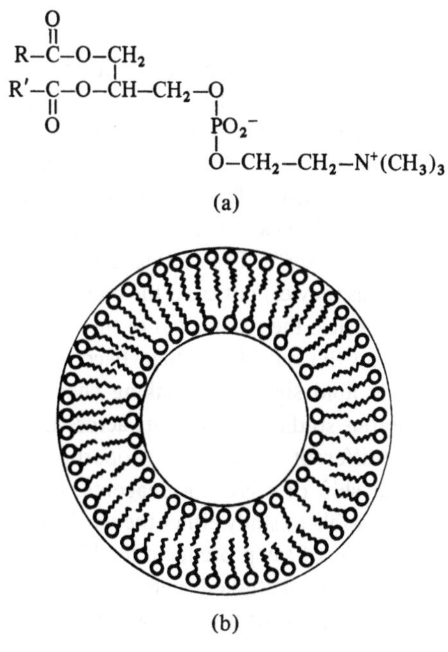

Fig. 12 a, b. Phosphatydyl choline vesicles. The formula of the lipid is at the top, where R and R′ are long chain alkyl groups, and the structure of the vesicle is shown below

The paramagnetic shifts and relaxation changes induced by Ln(III) on resonances can be used in a further way – to study binding constants. The binding of a cation to a bilayer surface is not the same as the binding to a monomer headgroup in free solution since the bilayer carries charge. The charge on the bilayer alters with Ln(III) binding and the binding is then anti-cooperative. Thus the initial binding constant, K, of Ln(III) to the headgroups of lecithin is larger than that to GPC, the isolated headgroup. Chelation is probable in the bilayer but the binding constant falls with occupancy as the charge builds up

$$K_x = Ke^{-xC}$$

where x is the fractional occupancy and C is a constant.

This type of binding is likely to be important in biology and has been analysed in somewhat different ways by McLaughlin[65)], Hauser[66)], and Ehrenberg[67)].

In the earlier sections we have also stressed the nature of hydration of lanthanide ions as revealed by changes in shift ratios and τ_s. It is extremely interesting in this respect that Ln(III) ions as a series do not bind lecithin head-groups in the bilayer in the same way as they bind the isolated head group, glycerol-phospho-choline (GPC) in free aqueous solution. We see this clearly in that all shift probes, i.e. including Tm(III), give the same shift ratios on binding the lipid bilayer but they give different shift ratios from those with GPC. The structure of the Ln(III) head group is different even for the axial probe Pr(III). Now it is also found that whereas the shift ratios are constant i.e. independent of Ln(III) for the head groups of the bilayer they change greatly for GPC in the second as opposed to the first part of the series. We take this to mean that the hydration of the cations is different in the head group from that at the membrane surface.

The biological significance of these studies is that they demonstrate that the effect of say calcium ions on membrane structures could be significant. Now in biology calcium ions are distributed unevenly being at $< 10^{-7}$ M inside cell and $> 10^{-3}$ M outside cells. Thus the binding is much more likely to the outside of the bilayer which could cause two effects

(a) It could stabilise particular lipids on the outside
(b) It could affect a difference in geometry of the lipid on the two sides of the membrane leading to differential binding of other reagents and even curvature of the bilayer.

It must be remembered the biological lipids are of many varieties some of which - the negatively charged lipids, can bind calcium strongly.

The Ln(III) cations can be used in another fashion in the study of vesicles as pointed out by Bystrov[62)]. The cations can be contained either only inside or only outside the bilayer as they cannot cross the membrane. Using a shift probe, say Pr(III), on the inside of the vesicle the resonances of the head-group of the inside lipids are shifted relative to those on the outside. It then becomes possible to follow differentially the inside and outside lipid signals on making changes in the bathing solution. This has permitted numerous studies of transport across the lipid bilayer.

A different study of the properties of the lipid layer can be made by incorporating organic shift or relaxation reagents, Ln(III)(FOD)$_3$, in the membrane[53)] (Fig. 13). It is then possible to study either the partitioning of molecules into the bilayer, e.g. $CHCl_3$, or the binding of say proteins to the bilayer surface. While we have seen both these effects we have not yet studied them in detail but hopefully this procedure will be used to find the special parts of proteins which bind to membranes.

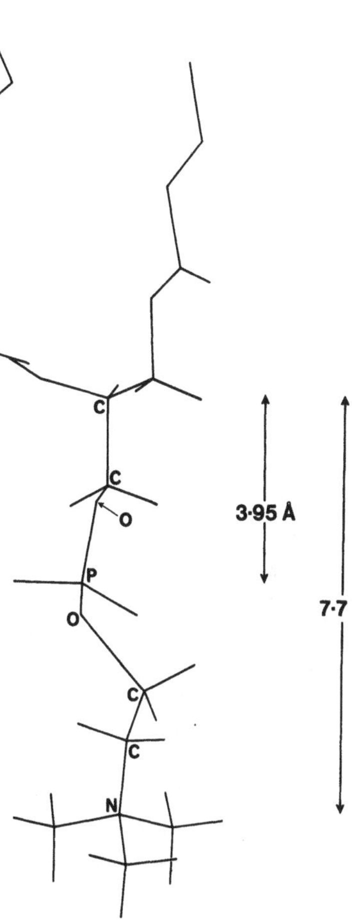

Fig. 13. The structure of the lipid in the vesicle as derived from Ln(III) shift and relaxation data

b. Proteins

Whereas no clear idea of the role of cations in the structure and function of biological membranes has emerged as yet the roles of cations such as Mg^{2+} and Ca^{2+} in proteins has been very clearly shown. Magnesium ions are often involved in the form of complexes with small phosphate-carrying molecules e.g. adenosine triphosphate, ATP. On the other hand calcium ions bind to proteins directly, frequently acting in a conformational trigger, linking an extra-cellular event with the onset of a mechanical or chemical intracellular response. Earlier in this article I made passing reference to the use of Ln(III) cations in the study of a kinase i.e. an enzyme which transfers phosphate from ATP to phosphoglycerate, see Fig. 5. The basic methods for such a study were worked out while analysing the binding of metal ions to lysozyme – a quite different protein. In effect the structure of a protein can be mapped radially from Gd(III) in a conventional experiment, here using Gd(III) · ATP, or by using relaxation measurements on resonances shifted by other paramagnetic probes e.g. Pr(III) · ATP. More detailed mapping is possible using the shift and relaxation data together. It is also possible of course to map

Table 12. Some proteins studied by Ln(III) reagents

Protein	Study
Lysozyme	Ln(III) probes established structure in solution
Neurotoxins	Uncovering of hydrophobic surface
Protease inhibitors	Assignment of resonances
Cytochrome-c	Assignment of resonances
Transferrin[55]	Ln(III) probes established the Fe(III) binding sites
Calmodulins	Uncovering of Ca(II) binding sites.

from different binding sites using different probes. Thus $[Cr(CN)_6]^{3-}$ binds at quite a different position from that favoured by Ln(III). In the study of lysozyme and cytochrome *c* a variety of shift and relaxation reagents have been used but here we are concerned with lanthanides alone (Table 12).

Both in the binding of Ln(III) · ATP, or Ln(III) · pyrophosphate to phosphoglycerate kinase, and in the binding of Ln(III) to lysozyme, the N.M.R. study was one in which the Ln(III) ion was in fast exchange, on and off the protein surface. The on/off rates of Ln(III) from the proteins which bind calcium more tightly is slow. This situation has been put to good use by Sykes and his co-workers[22)] in the examination of parvalbumin. They studied the protein fully bound with the shift probe Yb(III) using the Curie-spin relaxation and studying relaxation at two field strengths, see Eq. 4. Their results show that parvalbumin has a similar structure in solution to that found in crystals.

In the study of Ln(III) binding to lysozyme a major point of interest is the differences observed in the shift ratios with different Ln(III) cations. Once again the ratios for the second part of the series are very different from the first part. The situation is comparable with that in Dobson's study of the rigid frame dicarboxylate, Ref. 29. Figure 14 shows the centre of lysozyme indicating Glu 35 and Asp 52 the two carboxylate groups which assist the binding. They are too far apart to bind simultaneously except through intermediate water molecules. Although the site is roughly a rigid frame metal-binding is complicated by the possibilities of (a) different hydrates (b) different metal sites. In fact X-ray diffraction studies show that there is occupancy of at least two different sites in crystals. The N.M.R. data using Gd(III) binding to lysozyme in solution do not indicate that there is more than one heavily preferred site. Despite great effort there is no clear description of the binding of Ln(III) ions to this site especially since use of La(III) and Lu(III), diamagnetic N.M.R. probe reagents, show that there is a significant, though not large, conformation change when binding occurs.

The major important point to notice is undoubtedly the role of hydration associated with metal ions in proteins. The hydration is very clearly very dependent upon small changes in "ion size" much as was observed for the small model ligands.

Dobson and co-workers extended the N.M.R. study of carboxylate complexes to a rigid-frame dicarboxylate[29)] and showed that in this case the shift ratios were again different for the two halves of the Ln(III) series. They proposed that although the structures were similar for Pr(III) and Tm(III) the former behaved as an axial and the latter as a rhombic probe. Against the background of the other studies on non-chelating ligands and on dipicolinate we can extend the discussion. It now is likely that the major

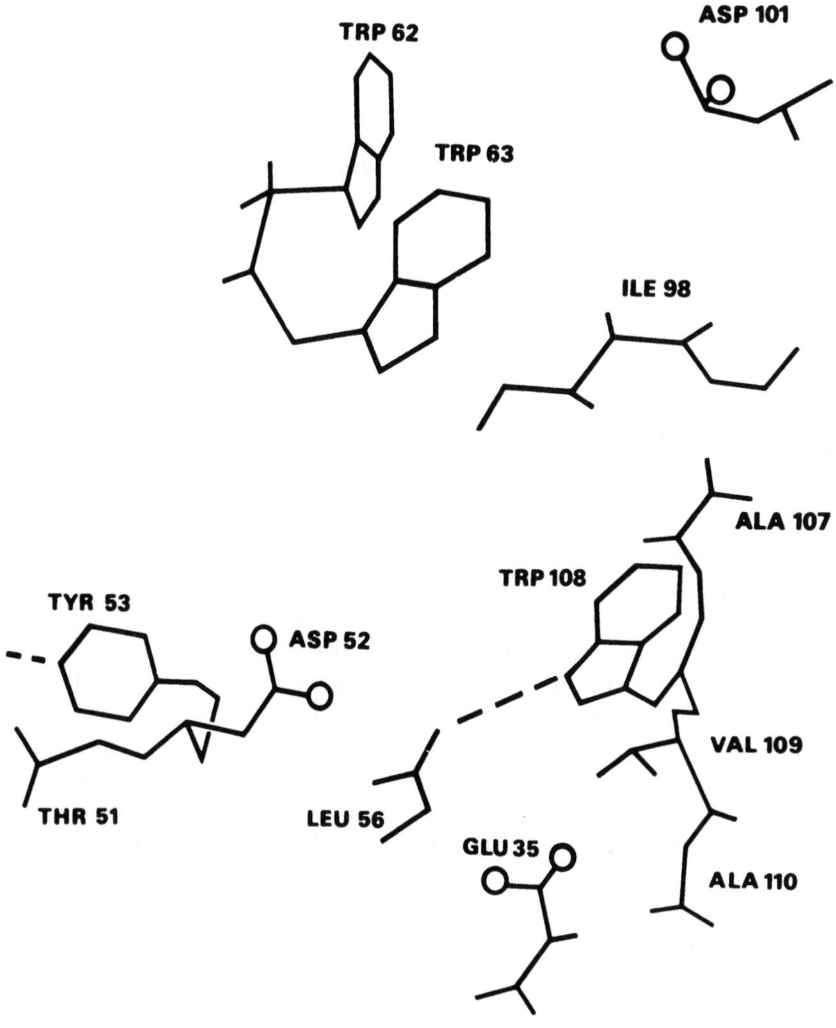

Fig. 14. Active site region of lysozyme. The Ln(III), NMR study demonstrated the structure in solution and showed that Val 109 flips rapidly, that Trp 108 undergoes conformational changes, that Trp 62 flips and that Tyr 53 flips rapidly

difference between the Ln(III) complexes is in the structures of the *hydration* of the complexes. This will cause minor differences in ligand-metal geometry but can cause quite major effects on the N.M.R. parameters. As in all other cases Dobson et al.[29] went on to satisfy themselves that the N.M.R. data for the Pr(III) complexes could be simply interpreted in terms of the *known* rigid-frame structure of the ligand assuming axial symmetry while the data for the Tm(III) ion could not be interpreted in this way.

Looking at all the data on N.M.R. shifts in Ln(III) complexes which we have collected we draw the following conclusion. There is often but not always a change in the structure of Ln(III) series of complexes. The Ln(III) ion at which a structural change occurs is not predictable, but it is probable that for many Ln(III) complexes there is an equilibrium mixture of structures. (The anomalies in shift ratios appear for different Ln(III) complexes with different ligands.)

IV. The Lanthanide Ions and Mobility in Biology

The function of metal ions such as iron in biology does not concern gross movement of the atom from its coordination site. Such movement would involve constant unfolding and folding of the protein since the sites are deeply buried. These cations do not exchange at appreciable rates. This is not to say that substrate exchange and small ligand rearrangements are not anything but fast. The latter is a requirement of good catalysis. Now ions such as calcium and magnesium have quite different functions and quite different mobilities on and off their binding ligands and they bind to surface loops of proteins not in deep grooves. How can we study surface mobility of say calcium(II)? This is an important problem which has not been solved.

It might be asked why it is necessary to study a protein in such detail. Figure 15 illustrates the problem. On combination of calium ions with calcium binding proteins there is a conformational change. Moreover the proteins are sensitive to environments, salt and temperature. Crystallographic methods will give the major features of these proteins but not the changes in structure nor the dynamics. Even the structures are biased by the salt media of crystallisation. Our objective is to define the structural responses of proteins and their dynamics in solution. This applies to the oxidation states changes of cytochrome *c*, to the conformational changes of lysozyme and to the fluctuating structures of calmodulins amongst many others. It implies equally to lipid, polynucleotide or polysaccharide structures. Now the study of structure in solution is difficult and of the dynamics of structures more difficult still. NMR is the most powerful method but it needs

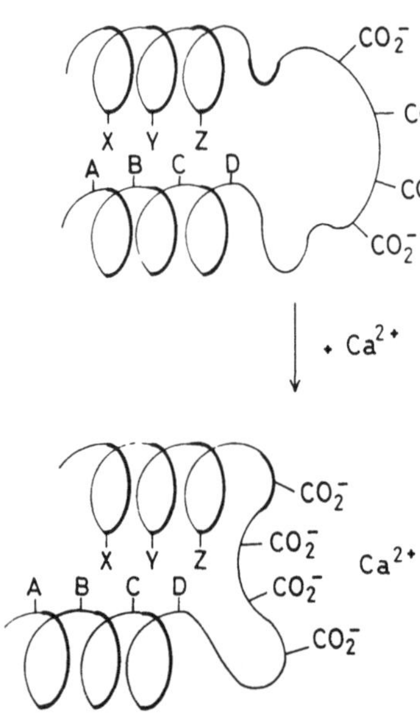

Fig. 15. A schematic representation of the binding site region of calcium trigger proteins (after Kretsinger). Using Ln(III) ions Sykes and coworkers[18] demonstrated that the solution and crystal structures were very similar

every auxilliary tool to assist it. The Ln(III) shift and relaxation analysis is one such procedure. The formidable difficulties which face anyone wishing to analyse observations in biological systems are listed in the next paragraphs with special reference to the case in which Ln(III) methods are most valuable – the study of Ca(II) activity. It should not escape the attention of chemists that Ca(II) ions are of the utmost importance in biological signalling e.g. at nerve junctions, in muscle action, in cell division, in fertilisation, in hormone control, in energy control and so on.

V. Biological Effects of Lanthanides

The following concluding remarks can be made about the structures, thermodynamics and kinetics of complex ion reactions in biology where the energies are purely electrostatic and as revealed by Ln(III) studies: There are many tens of such studies.

(1) Even in isostructural series a maximum in stability or rate constants can occur almost anywhere as radius decreases but changes are likely to be small.
(2) For series in which structure varies with "radius" more pronounced changes can occur for small changes in "radius".
(3) The temperature dependence of stability is often more sensitive than the stability itself to changes in "radius".
(4) The mobility of ligands on the surface of an ion and on-and-off the ion are very dependent on "radius".

Biological function of an ion can be analysed in the following manner

$$\text{Effect} \cdot \alpha \quad [\text{M}] \cdot \text{K}_{\text{aq}} \cdot \varrho \cdot \text{F(structure)} \qquad (10)$$

where K_{aq} is the formation constant of the complex and ϱ is a transfer (partition) coefficient for its movement into the proper solvent environment. F(structure) describes the free energy required to get the metal complex through the structural changes associated with activity and causing a known effect. Let us assume [M] is fixed, $K_{aq}\varrho$ will show all the variability mentioned above but most interestingly to our understanding are changes in F(structure) since very small changes in radius may affect it grossly. Thus although it may not be obvious why Ca^{2+} behaves so differently from Mg^{2+} and even from Sr^{2+} in biological systems, possible explanations exist which can be tested.

We can therefore look for biological effects of the Ln(III) series in two senses (a) How do the NMR parameters (or parameters of other physical methods) change with Ln(III), (b) How do biological changes vary with Ln(III). Whether we can interpret the observations or not they will certainly lead to an increased appreciation of the nature of electrostatic interactions. For those who wish to look at the complexity of the problem I include Ref. 69 onwards.

I. References

1. Spitsyn, V. I., Martynenko, L. I.: Bull. Acad. Sci. U.S.S.R. Div. Chem. Sci. *23,* 940 (1974)
2. Yatsimirskii, K. B., Davidenko, N. K.: Coord. Chem. Reviews *27,* 223 (1979)
3. Williams, R. J. P.: Proc. 4. Intl. Congr. Pharmacology, Vol. V, 227, 1970

4. Williams, R. J. P.: Quart. Rev. Chem. Soc. (London) *24,* 331 (1970)
5. Darnall, D. W., Birnbaum, R.: J. Biol. Chem. *245,* 6484 (1970)
6. Hinckley, C. C.: J. Amer. Chem. Soc. *91,* 5160 (1969)
7. Saunders, J. K. M., Williams, D. H.: Chem. Commun. Chem. Soc. (London), 422 (1970)
8. Barry, C. D., Hill, H. A. O., Sadler, P. J., Williams, R. J. P.: Proc. Roy. Soc. (London) *A, 334,* 493 (1973) and references therein
9. Barry, C. D., Glasel, J. A., North, A. C. T., Williams, R. J. P., Xavier, A. V.: Nature, *232,* 236, 1971
10. Nieboer, E.: Structure and Bonding *22,* 1 (1975)
11. Reuben, J.: Progr. Nucl. Mag. Res. Spectroscopy *9,* 1 (1973); Reuben, J., Elgavish, G. A. in: Handbook on the Physics and Chemistry of Rare Earths (ed. K. A. Gschneider, L. Eyring), North Holand, Amsterdam, Vol. 4 (1979), p. 483
12. Dobson, C. M., Levine, B. A. in: New Techniques in Biophysics and Cell Biology (eds. R. Pain, B. Smith), Vol. *3,* Chap. 2, Wiley, New York, p. 19. (This article gives a summary of work up to 1975)
13. Inagaki, F., Miyazawa, T.: Progr. N.M.R. Spectroscopy *67,* 14 (1981)
14. Barry, C. D., Dobson, C. M., Ford, L. O., Sweigart, D. A., Williams, R. J. P. in: Nuclear Magnetic Shift Reagents (ed. R. E. Sievers), Acad. Press, New York, 1973, p. 173
15. Dobson, C. M., Williams, R. J. P., Xavier, A. V.: J. Chem. Soc. Dalton 2662 (1973)
16. Reilly, C. N., Good, B. W.: Anal. Chem. *47,* 2110 (1975)
17. Bleaney, B.: J. Mag. Res. *8,* 91 (1972)
18. Bleaney, B., Dobson, C. M., Levine, B. A., Martin, R. B., Williams, R. J. P., Xavier, A. V.: Chem. Commun. Chem. Soc. London, 791 (1972)
19. Gueron, M.: J. Magnet. Res. *19,* 58 (1975)
20. Vega, A. J., Fiat, D.: Molec. Phys. *31,* 347 (1976)
21. Fung, L. W.-M., Ho, C.: J. Amer. Chem. Soc. *31,* 347 (1977)
22. Lee, L., Sykes, B. D.: Biochemistry *19,* 3208 (1980)
23. Barry, C. D., North, A. C. T., Glasel, J. A., Williams, R. J. P., Xavier, A. V.: Nature *236,* 232 (1971)
24. Swaninathan, V., Sundaralingam, M.: CRC Critical Rev. Biochem. *6,* 245 (1979)
25. Dobson, C. M., Geraldes, C. F. G. C., Ratcliffe, G., Williams, R. J. P.: Eur. J. Biochem. *88,* 259 (1978)
26. Tanswell, P., Westhead, E. W., Williams, R. J. P.: ibid. *63,* 249 (1976)
27. Blake, C. C. F., Evans, P. R.: J. Molec. Biol. *84,* 585 (1974)
28. Levine, B. A., Williams, R. J. P.: Proc. Royal Soc. London *345 A,* 1640 (1975)
29. Delepierrie, M., Dobson, C. M., Menear, S. L.: J. Chem. Soc. (Dalton) 678 (1981)
30. Dezube, B., Dobson, C. M., Teague, C. E.: (A study of Lu(III) tryptophan complexes) to be published
31. Alsaadi, B. M., Rossotti, F. J. C., Williams, R. J. P.: J. Chem. Soc. (Dalton) 597 (1980)
32. Hauser, H., Levine, B. A., Phillips, M. C., Williams, R. J. P.: Nature *261,* 390 (1976)
33. Hauser, H., Levine, B. A., Williams, R. J. P.: to be published
34. Alsaadi, B. M., Rossotti, F. J. C., Williams, R. J. P.: J. Chem. Soc. (Dalton) 813 (1980)
35. Alsaadi, B. M., Rossotti, F. J. C., Williams, R. J. P.: ibid. 2147 (1980)
36. Alsaadi, B. M., Rossotti, F. J. C., Williams, R. J. P.: ibid. 2151 (1980)
37. Grenthe, I.: Kemia-Kemi, *6,* 234 (1978) and references therein
38. Habenschuss, A., Spedding, F. H.: J. Chem. Phys. *442, 73* (1980)
39. Reuben, J., Fiat, D.: J. Chem. Phys. *51,* 4909 (1969)
40. Stability Constants Chem. Soc. (London) publication no. 17 (eds. L. G. Sillen and A. E. Martell), 1964
41. Phillips, C. G. S., Williams, R. J. P.: Inorganice Chemistry, Oxford Univ. Press, Oxford (1966)
42. Dzhurinskii, B. F.: Russian J. Inorg. Chem. (in English) *41,* 25 (1980) and references therein discussing "Tetrad Effects"
43. Cummings, S., Enderby, J. E.[1], Howe, R. A.: J. Phys. Chem. Solid State Physics *1,* 13 (1980)

---

[1] Prof. J. E. Enderby has informed me that the calcium ion hydrate which is $[Ca(H_2O)_6]^{2+}$ in strong solution may well be of higher coordination number, 8–10, in dilute solutions as determined by neutron diffraction studies

44. Grenthe, I.: Acta Chem. Scand. *18,* 293 (1964)
45. Grenthe, I.: pemia-Kemi[37)]
46. Deibler, H., Eigen, M.: Z. Phys. Chem. *20,* 229 (1959)
47. Williams, R. J. P. in: Biological Membranes (ed. D. S. Parsons), Oxford Univ. Press, Oxford (1975), p. 106
48. Poona, N. S., Bajaj, A. V.: Chem. Rev. *79,* 389 (1979)
49. Chakrabarti, P., Venkatesan, K., Rao, C. N. R.: Proc. Roy. Soc. (London) *A, 375,* 127 (1981)
50. Horrocks, W. De W., Sudnick, D. R.: Science *206,* 1194 (1979)
51. Horrocks, W. De W., Sudnick, D. R.: J. Amer. Chem. Soc. *101,* 334 (1979)
52. Horrocks, W. De W., Homquist, B., Vallee, B. L.: Proc. Natl. Acad. Sci. (U.S.) *72,* 4764 (1975)
53. Mann, S., Williams, R. J. P. to be published
54. Dobson, C. M., Williams, R. J. P. in: Metal-ligand Interactions in Organic Chemistry and Biochemistry, Part I (eds. B. Pullmann and N. Goldblum), Reidel Pub. Co., DordrechtHolland (1977), 255
55. Alsaadi, B., R. C. Woodworth, Williams, R. J. P.: Inorganic Biochemistry, *15,* 1 (1981)
56. Hoyle, N.: D. Phil. Thesis, Oxford University
57. Desreux, J.: Inorg. Chem. *19,* 1319 (1980)
58. Williams, R. J. P.: The Analyst *78,* 586 (1953)
59. Pedersen, C. J.: J. Amer. Chem. Soc. *92,* 386 (1970)
60. Lehn, J. M.: Structure and Bonding *16,* 1 (1973)
61. Williams, R. J. P. in: Proc. Fourth Internatl. Conf. Pharmacology, Vol. V, p. 227 (1970)
62. Bystrov, V. F., Dubrovina, N. I., Bavsukov, L. T., Bergelson, L. D.: Chem. Phys. Lipids *6,* 343 (1971)
63. Hauser, H., Phillips, M. C., Levine, B. A., Williams, R. J. P.: Nature *261,* 390 (1976)
64. Brown, M. F., Seelig, J.: Nature *269,* 721 (1977)
65. Büldt, G., Seelig, J.: Nature *271,* 182 (1978)
66. Hauser, H., Levine, B. A., Williams, R. J. P.: Biochim. Biophys. Acta *508,* 450 (1978)
67. Westman, J., Eriksson, L. E. G.: Biochim. Biophys. Acta *557,* 62 (1979)
68. dos Remedios: Nature *270,* 750 (1977)
69. Crompton, M., Heid, I., Baschera, C., Carafoli, E.: FEBS Letters *104,* 352 (1979)
70. Lenkinsi, R. E., Stephens, R. L.: J. Inorg. Biochem. *15,* 95 (1981)
71. Brewer, J. M., Carreira, L. A., Irwin, R. M., Elliot, J. I.: J. Inorg. Biochem. *14,* 33 (1981)

Uranyl Photophysics

Christian K. Jørgensen[1] **and Renata Reisfeld**[2,*]

1 Département de Chimie minérale, analytique et appliquée, Université de Genève, CH 1211 Geneva 4, Switzerland, and

2 Department of Inorganic and Analytical Chemistry, and the Energy Center, Hebrew University, Jerusalem, Israel

The electron transfer bands of the triatomic uranyl group are due to a complicated manifold of excited levels described by relativistic effects in linear symmetry. The lowest of these levels can live for longer than 1 millisecond (in the absence of adjacent hydrogen atoms) and has its own chemistry, reacting with many anions, univalent silver and thallium, in a way entirely different from the groundstate. It is nearly as oxidizing as free fluorine; forms exciplexes; and abstracts hydrogen atoms from organic molecules (perhaps also in the ocean) and from water. The spectroscopic and luminescent properties are compared with the hexafluorides and the dioxo complexes of quinquevalent and sexivalent uranium, neptunium, plutonium and americium, and with roughly octahedral species, and tungstates. Uranyl glasses are useful for energy transfer to lanthanides and converting solar energy via photovoltaïcs.

* Enrique Berman professor of solar energy

Structure and Bonding 50

1 Comparison Between the Chemical Properties of Other Elements and the Uranyl Groundstate

Among the elements known before transuranium elements started to be synthesized in 1940, uranium has a unique characteristic, the extreme stability of the triatomic *uranyl ion* OUO^{+2}. Not only are the numbers of uranium(VI) compounds larger than of U(IV), and far larger than of the two other oxidation states U(V) and U(III) known from non-metallic compounds, but until the preparation of UOF_5^- discussed below, the only two U(VI) compounds known to contain less than two oxygen atoms per uranium atom were the octahedral molecules UF_6 and UCl_6.

Before it was realized that the atomic number Z corresponds to the positive charge Ze of the nucleus, and before atomic weights were directly determined from mass spectra, it was very difficult to obtain autonomous chemical evidence for the elemental character of a solid[1]. Klaproth characterized in 1789 the uranyl salts such as chloride, nitrate, sulphate, . . . and obtained brown UO_2 by strong reduction of UO_3 by hydrogen at red heat. UO_2 was considered to be an almost metallic element (much like tellurium) until Peligot in 1841 prepared anhydrous green UCl_4 by heating a mixture of the oxide and finely divided carbon in a stream of chlorine (in analogy to Ørsted preparing anhydrous aluminium chloride). The subsequent reduction of UCl_4 with metallic sodium provided a black, pyrophoric, metallic powder in such a yield that it became clear that uranyl salts and UO_2 contain constitutional oxygen.

In crystalline structures[2] UO_2^{+2} is linear with remarkably short U–O distances 1.7 to 1.8 Å, whereas water molecules and oxygen-containing anions (such as nitrate) bound in the equatorial plane typically have U–O some 0.7 Å longer. The most frequent number of neighbour atoms in the equatorial plane is 6, but other numbers 5 and 4 also occur, in particular voluminous halide anions such as $UO_2Cl_4^{-2}$ and $UO_2Br_4^{-2}$.

In aqueous solution, the uranyl ion is distinctly more acidic (forming hydroxo complexes) than magnesium, but comparable to other bivalent elements such as cadmium and lead. The deprotonation of aqua ions to hydroxo and oxo complexes depends in general[3] on four parameters. The most obvious is pH varying in aqueous solution roughly between -1 and $+15$. Solutions more acidic than -2 or more alkaline than $+16$ usually have so low water activity that they can hardly be characterized as aqueous solutions[4]. The general tendency for deprotonation is to increase not only as a function of increasing pH, but also as a function of increasing oxidation state z. Thus, protonation of ClO_4^- and of OsO_4 cannot be detected in acidic aqueous solution. However, besides pH and z, there are two other factors of some influence. In relatively electrovalent cases, there is a marked tendency for decreasing ionic radii to enhance deprotonation in the direction aqua → hydroxo → oxo ligands. Without necessarily having a quantitative expression (say, the ratio between z and the ionic radius) this tendency corresponds to the "diagonal similarities" in the Periodic Table between elements such as beryllium(II) and aluminium(III); boron(III) and silicon(IV); or carbon(IV) and phosphorus(V). Even if we restrict attention to aqua ions, it is evident that $Be(OH_2)_4^{+2}$ is far stronger as a Brønsted acid than $Mg(OH_2)_6^{+2}$ and the aqua ions of the larger alkaline-earths; that the hexaqua ions of the small Al(III), Cr(III) and Ga(III) are stronger acids than the aqua ions of the trivalent lanthanides[5,6]; and the weakest acidity of any known M(IV) aqua ion is exhibited by thorium(IV) having the largest ionic radius. However, there is definitely a fourth parameter prevailing, of less well-defined origin, of enhanced acidity due to stronger covalent bonding or more pronounced chemical polarizability[7]. We already mentioned that Cd(II) and Pb(II) are more acidic than Mg(II), in spite of their larger ionic radii; it is well-known that copper(II) and iron(III) are more acidic than comparable ions, and the aqua ions acquire pK values as low as 1 in the extreme cases of palladium(II), mercury(II) and thallium(III). It is perhaps not too surprising that preferential covalent bonding in hydroxo and oxo complexes decreases the pK values of aqua ions, though it cannot be argued that the aqua ions are destabilized[7–10] since several of the cases mentioned have unexpected high hydration energies of the gaseous ions M^{+z}. It seems that a low coordination number *N* under equal circumstances enhances the Brønsted acidity as very strikingly seen[11, 12] in the case of gold(III) compared with the higher oxidation state platinum(IV). An extreme situation is found in univalent halogens, where the deprotonation of hypochlorous acid HOCl and of HOBr shows pK values comparable to aqua ions of trivalent lanthanides or to diprotonated phosphate $(HO)_2PO_2^-$.

The uranyl ion does not fit this description of deprotonation as a function of four parameters, because of the simultaneous presence of oxo and aqua ligands. One would then have expected a tautomeric rearrangement to a hydroxo complex such as $U(OH)_4^{+2}$. Actually, very few of the -yl cations of old text-books have survived; they seem to be oligomeric hydroxo complexes such as the zirconyl and bismuthyl ions $Zr_8(OH)_{16}^{+16}$ and $Bi_6(OH)_{12}^{+6}$. However, the recent studies of visible absorption spectra[13] have confirmed the blue vanadyl ion VO^{+2} as a monomer lacking proton affinity in strong perchloric acid, and it seems[14] that it has four weakly bound water molecules $OV(OH_2)_4^{+2}$ with the symmetry of a tetragonal pyramid. Both kinetic (^{18}O exchange rate) and spectrophotometric studies of the uranyl ion in aqueous acids confirm even more firmly the lack of any detectable proton affinity and of any accessible path of tautomeric proton mobility. There are very few dioxo complexes known of comparable stability, the closest analogy perhaps being CO_2. Though the idea of multiple bond-order outside the group of carbon compounds seems[15] rather devoid of experimentally verifiable consequences (e.g. in OPX_3 and in PO_4^{-3}) though there may be a justification for classifying the short X–O bonds lacking detectable proton affinity in CO (being isoelectronic with N_2) and in the vanadyl and uranyl ions as triple bonds.

Before 1945, the general consensus was that uranium is a homolog[16, 17] to the column chromium, molybdenum and tungsten. Whereas yellow CrO_4^{-2} protonate to orange $HOCrO_3^-$ in rapid equilibrium with the dichromate $O_3CrOCrO_3^{-2}$ analogous to pyrosulphate[18] there is no evidence for CrO_3 forming cations in acidic solution. By the same token, MoO_3 and WO_3 only form colloidal solutions in perchloric acid, whereas acids containing complexing anions (such as chloride) may form mixed oxo-chloro complexes such as known[19] of niobium(V) and tantalum(V). The quiet acceptance of uranyl salts may have been an etymologic misunderstanding (much like acetates denote both ester molecules and ionic salts) since chromyl chloride CrO_2Cl_2 consists of tetrahedral molecules closely similar to sulphuryl chloride. It was first pointed out by Cotton[20] that crystal structures of dioxo molybdenum(VI) compounds (containing other anions) generally are *cis*-octahedral MoO_2X_4 and not *trans*-octahedral, as would be the case for uranyl compounds. This fact has been further discussed[21]. Actually, the closest analogs[1] to the uranyl ion are *trans*-octahedral dioxo complexes formed by the $4d^2$ system ruthenium(VI) and the $5d^2$ systems rhenium(V) and osmium(VI), $ReO_2(NH_3)_4^+$ (having measurable proton affinity in acidic solution) and purple, diamagnetic $OsO_2(OH)_4^{-2}$ (to be compared with orange, tetrahedral ruthenate RuO_4^{-2} with triplet groundstate). It is a common characteristic of these dioxo complexes that they have both their d-like electrons in the same non-bonding orbital with the angular function proportional to *xy* (with all six ligating atoms on the Cartesian axes, among which the oxygen nuclei on the *z*-axis). It is by no means evident that the specific stereochemistry of the uranyl ion has anything to do with the empty 5 f shell, and the major part of the covalent bonding seems to be assured by the empty 6 d orbitals in L.C.A.O. models.

The uranyl ion is exceptional in another way. General experience from organic chemistry would suggest that the oxo ligands might be substituted by the isoelectronic groups NH^{-2} or CH_2^{-2}. We are not aware of any such substituted uranyl ions, though the nitrido complex of osmium(VIII) $NOsO_3^-$ is known[18] to lack proton affinity at pH = 0. It may be that NUO^+ would be easier to prepare; the solid nitride UN_2 crystallizes in fluorite lattice. Again, sulphur-substituted SUO^{+2} has apparently not been reported. When solutions of H_2S and HS^- were important reagents for qualitative analysis, it was already well known that precipitates "uranium red" of ill-defined stoichiometry could be obtained. It is likely that they have sulphur atoms bound in the equatorial plane, when compared to the dithiocarbamate complexes $UO_2(S_2CNR_2)_2$ and $UO_2(S_2CNR_2)_3^-$ having a fox-red colour due to the electron transfer bands to be further discussed below.

Both crystalline salts and glasses containing uranyl ions frequently show strong greenish yellow luminescence[22] first described by Brewster in 1833. It is intrinsic, and not due to trace activators. The shift between the absorption bands in the blue and the emission bands in the yellow (having one narrow band in the green in common) formed the basis for the law formulated by Stokes in 1852. Both the absorption and the emission spectra consist of approximately equidistant progressions, which were later compared with the vibrational spectra of diatomic molecules (having only one characteristic vibrational frequency for each electronic state). However, it is beyond doubt that this comparison is over-simplified; as we shall see below, there is a large number of adjacent electronic states above the Stokes threshold at 20 500 cm^{-1} (488 nm) and it seems to be a numerical accident[23, 24] that several vibrational progressions belonging to differing electronic states can be approximately combined to one progression in the absorption spectrum. Further on, it is a group-theoretical problem that if the uranyl ion in the relatively long-lived excited state remains strictly linear, the emission would only be allowed as electric dipolar transitions, if one (or another odd number) vibrational frequency corresponding to the antisymmetric nuclear motion having odd (ungerade) parity is co-excited. We return to several of these problems below, but it is well established that the equidistant progression in the emission spectrum has higher wave-number differences (close to 850 to be compared with 710 cm^{-1}) corroborating the expectation that the large Stokes shift is connected with a much longer equilibrium U-O distance in the excited state than in the groundstate. In other cases, such as tetrahedral and octahedral manganese(II) chromophores[25] the large Stokes shift corresponds to strongly contracted equilibrium internuclear distances in the fluorescent state. However, a further contraction would not be conceivable in the uranyl ion, but it is interesting to note how relatively unimportant the ligands in the equatorial plane are for the positions of the vibrational components in the emission spectrum. In this sense, the chemical arguments for an independent triatomic uranyl ion in condensed matter are strongly supported by spectroscopic evidence, though exceptions occur, such as UO_6^{-6} discussed in Sect. 8 and certain uranyl glasses in Sect. 9.

Quite independent of past and future applications of uranyl luminescence[22, 25] the phenomenon had an enormous impact on human history in 1896. As kindly pointed out by Professor Hurwic to one of us[17], the mathematician Henri Poincaré suggested that since the X-rays (emitted by Crookes tubes and discovered by Röntgen) make $BaPt(CN)_4$, $4\,H_2O$ fluoresce, it might perhaps be that some fluorescent materials simultaneously emit X-rays. Comptes-Rendus (of Paris Academy of Sciences) for January and February 1896 contain several negative reports. However, Henri Becquerel (who had worked extensively with spectroscopic properties of uranyl salts) found that they emit indeed X-rays. A few weeks later, he found that the emission of X-rays also occurs in weak light (on a dull day) and even without any accompanying luminescence at all. This observation of radioactivity (followed by the discovery of radium by Maria Sklodowska and Pierre Curie in 1898) is the basis for the recognition of nuclei in atoms. This was such an unexpected idea that it might easily have gone unnoticed for several generations, whereas concepts of comparable importance, such as atomic spectral analysis or X-ray spectrometry, were likely to be discovered one year or another. It is worthwhile noting with Pais[26] that the origin of the radioactive energy became an ardent problem several years before the interpretation in terms of spontaneous transmutation of elements heavier than bismuth became an acceptable rationalization. At this point, it may be useful to remember the importance of what Glashow[27, 243] calls "passive experimentation". It is conceivable that careful investigation of definite minerals[28, 29] may allow the detection of species containing unsaturated quarks or very heavy "elementary" particles protected against rapid decay by strong selection rules.

2 Electron Transfer Spectra

Many text-books argue that transitions localized inside a partly filled shell of transition-group chromophores have been rationalized much earlier and to a detailed extent, compared with the qualitative understanding of electron transfer spectra. It is true that the transitions inside the 3 d, 4 d or 5 d shell (in the iron, palladium and platinum groups, respectively) are described by "ligand field" theory[30–32] where the one-electron energy differences (originally ascribed to the minute deviations of the Madelung potential from spherical symmetry) since 1956 have turned out to represent the anti-bonding character of some of the d-like orbitals, relative to other, approximately non-bonding d-orbitals. In transthorium compounds containing a partly filled 5 f shell, and even more so in lanthanides with a partly filled 4 f shell[33, 34] the excited states are quite similar to J-levels of a monatomic entity. However, these two facts are only a part of the truth; already in 1919, Shibata photographed spectra in the ultra-violet ascribed to transfer of an electron from reducing ligands to an oxidizing central atom, and before 1930, Fromherz made an extensive study of such electron transfer spectra, which have been reviewed[35, 36]. It was early recognized that oxidizing central atoms such as $3\,d^5$ iron(III), $3\,d^6$ cobalt(III), $3\,d^9$ copper(II), $5\,d^5$ iridium(IV) and $5\,d^6$ platinum(IV) have electron transfer bands at much lower wave-numbers than closed-shell d^{10} systems surrounded by the same ligands. Nevertheless, hexachloro complexes of tin(IV), antimony(V) and lead(IV) show strong electron transfer bands[37] presumably due to a relatively low-lying, empty 5 s or 6 s orbital.

At the same time as the d-group spectroscopy developed rapidly because of the visible and ultra-violet spectrophotometers becoming readily available, a large material of electron transfer spectra were collected[35,37]. In 1961, the generalization of *optical electronegativity* was proposed at a meeting of the American Chemical Society held in Chicago, by writing the wave-number of the first strong electron transfer band

$$h\nu_{e.t.} = (30\,000\ \text{cm}^{-1})[x_{opt}(\text{X}) - x_{uncorr}(\text{M})] \tag{1}$$

proportional to the difference between the optical electronegativity of the most reducing ligand X bound to the central atom M, for which the uncorrected optical electronegativity x_{uncorr} has not yet been corrected for anti-bonding character of the d-like orbital receiving the electron in the excited state, nor for effects of spin-pairing energy expressed as definite coefficients multiplied by parameters of interelectronic repulsion[31,38]. Even before these two corrections, x_{uncorr} generally increases strongly with the oxidation state of M, and is, for instance, 2.3 for iridium(IV) but 2.7 for Ir(VI) known from gaseous IrF_6.

It should be emphasized that Eq.(1) does not pretend an average precision better than 3000 cm^{-1}. It turns out that the Pauling values can be ascribed to the halide ligands, $x_{opt} = 3.9$ for F^-, 3.0 for Cl^-, 2.8 for Br^- and 2.5 for I^-. Once a large set of halide electron transfer bands had been collected[36,39] it was possible to evaluate x_{opt} for other ligands, such as 3.5 for H_2O, 3.2 for SO_4^{-2}, 2.6 for nitrogen-bound NCS^-, 2.8 for N_3^- and 2.6 for several RS^-. The approximate agreement with Eq. (1) of spectral shifts, varying the reducing ligand, is now accepted as a convincing argument for electron transfer character, though it must be admitted that inter-shell transitions[31] such as $4f \rightarrow 5d$ in lanthanide M(II) and M(III) compounds, and $5f \rightarrow 6d$ in transthorium M(III) and M(IV) can show shifts with halide ligands X^- amounting to a-quarter or a-third of the shift predicted by Eq. (1). A similar situation[31,39,40] occurs in what seems to be $5s \rightarrow 5p$ transitions in tin(II), antimony(III) and tellurium(IV) complexes, or $6s \rightarrow 6p$ transitions in thallium(I), lead(II) and bismuth(III). However, in the latter case, it can be discussed[41] to what extent the *l*-values prevailing in spherical symmetry[32,42] are a good approximation. The other way round, the rather exceptional octahedral lanthanide complexes MCl_6^{-3} and MBr_6^{-3}[43] and MI_6^{-3}[44] show genuine electron transfer bands for M = Sm, Eu, Tm and Yb with a shift from chloride to iodide some 20 percent larger than predicted from Eq. (1). This may be explained[32] by the reducing character of the iodide ligands being enhanced by the pronounced ligand-ligand anti-bonding interactions in the M.O. loosing an electron in the excited state.

From a theoretical point of view[30] one expects deviations from Eq. (1) to occur for reasons already present in monatomic entities. The optical transition $a \rightarrow b$ does not depend only upon the difference between the ionization energies but also effects of interelectronic repulsion to be described to the first approximation by

$$h\nu \cong \text{I(a)} - \text{I(b)} + \text{J(b, b)} - \text{J(a, b)} \tag{2}$$

where the parameter J(x, y) is the electrostatic interaction between the charge densities of the orbital x and of the orbital y. If the orbital b is empty in the groundstate, I(b) is not directly observable, but is defined by the electron affinity of the orbital b being I(b) − J(b, b). The last term J(a, b) of Eq. (2) represents, colloquially speaking, the attraction between the hole created in the orbital a and the electron in the orbital b. It is more than sufficient to explain the higher wave-numbers of electron transfer taking place across a longer distance, as was first studied by Linhard in the case of the ion-pair formed between $Co(NH_3)_6^{+3}$ and I^- to be compared with the complex $Co(NH_3)_5I^{+2}$ having a direct cobalt-iodine contact[45]. We return below to the general case of anisotropic complexes in relation to Eq. (2).

It seems clear today that the majority of complexes agreeing reasonably well with Eq. (1) must represent some standard M–X distance (corresponding to roughly additive ionic radii) providing for only small (and perhaps, in part, compensated) variations of J(a, b) from Eq. (2). A quite unexpected feature of the optical electronegativities is that when the ionization energies of valence-shell M.O. became available from photo-electron spectra[46] the lowest ionization energy I of solid halides[47, 48] turned out to be about 1 eV below (3.7 eV) x_{opt} and of gaseous halide molecules[48–51] approximately 1 eV above this quantity, where 3.7 eV represents the constant 30 000 cm^{-1} of Eq. (1). This is particularly surprising because neither Pauling's scale nor the optical electronegativities were intended to have a definite zero-point.

Since uranium(VI) is member[34, 52] of the isoelectronic series $K = 86$ (like the noble gas radon) there is no serious doubt that the absorption bands in the visible and near ultra-violet[23] are due to electron transfer. When it has been stated at several occasions[2, 22] that they are rather intramolecular transitions, we do not argue that strong absorption bands (e.g. of methylene blue cations or fluoresceinate anions) cannot occur, which are not at all electron transfer bands in the sense used in inorganic chemistry. However, the only way in which such a situation might occur in the uranyl ion would involve a transition (a → b) between two M.O. which were *both* highly delocalized on the uranium and the two oxygen atoms. When we took this problem up again[53] it became clear that the transition goes to a set of orbitals very close to be non-bonding 5 f electrons, and the transition is so exceptionally weak that the M.O. loosing the electron must have odd (ungerade) parity. Though uranium 6 d and 7 s orbitals may contribute significantly (in L.C.A.O. models) to the occupied M.O. having even parity, there is little reason to expect extensive delocalization of the odd M.O. due to mixing with filled 6 p or empty 5 f and 7 p orbitals. Thus, it is at least as justified to speak about electron transfer bands of the uranyl ion as in typical inorganic d-group complexes. This does not prevent that the uranyl ion presents several specific problems, and in particular, it is rather incompatible with Eq. (1).

Besides the *J*-levels belonging to the configuration $4f^q$, it was readily accepted that inter-shell transitions to $4f^{q-1}5d$ can be observed in many lanthanide compounds. This is a close analogy to the behaviour of monatomic entities, and there was a lasting hesitation to accept electron transfer spectra, being a concept foreign to atomic spectroscopy. The first systematic studies[54] of electron transfer spectra of trivalent lanthanides were of the bromides dissolved in almost anhydrous ethanol. The species is likely to be solvated MBr^{+2} but the stoichiometry is far better established in the hexahalide complexes MX_6^{-3} [43, 44]. In both cases, the systematic increase of the wave-number of the first electron transfer band as a function of M(Eu < Yb < Sm < Tm < ...) corresponding to $q = 6$, 13, 5, 12, ... in the groundstate demonstrates beyond doubt electron transfer to the partly filled 4 f shell.

There was known an isolated case of electron transfer bands in the lanthanides in the form of the yellow to red colours of many cerium(IV) complexes, including[43] orange $CeCl_6^{-2}$ and purple (rapidly decomposing) $CeBr_6^{-2}$. However, these bands did not seem to constitute a perfect analogy to the uranyl ion, because they are several hundred times as intense. Since their intensity is comparable to the electron transfer bands of iridium(IV) complexes, one might imagine that the weak transition in the uranyl ion goes to 5f and the strong transition in Ce(IV) to 5d. However, this would be highly unlikely in view of the 4f → 5d transitions typically starting in cerium(III) close to 30000 cm^{-1} [33] and was entirely disproved by the 4f group M(III) electron transfer spectra (which tend to be 10 to 50 times stronger than the absorption bands of the uranyl ion in the blue).

Both the $5f^2$ system UI_6^{-2}, and the three $5f^3NpX_6^{-2}$ and $5f^4PuX_6^{-2}$ (with X = Cl, Br and I) have quite strong electron transfer spectra[55] though not as intense as 5f → 6d inter-shell transitions, and hence, it is not a general property of electron transfer to the 5f shell to be very weak. One may conclude that electron transfer takes place from even-parity M.O. in the octahedral hexahalides, undoubtedly from t_{1g} at lowest wave-numbers[32, 36, 39].

The mere existence of octahedral molecules UCl_6 in the crystal structure[56] of the dark coloured solid provides a severe upper limit of x_{opt} for U(VI) if Eq. (1) can be applied. Though the reports of the electron transfer spectra of this molecule are rather ambiguous, it seems to have the first absorption band at 15000 cm^{-1} corresponding to x_{uncorr} = 2.5. Good arguments[31, 32] can be given for the first electron transfer band of OsI_6^{-2} at 9400 cm^{-1} to represent an almost absolute lower limit against spontaneous reduction, in what case x_{uncorr} cannot be higher than 2.7 for U(VI). This forms a striking contrast to the (very weak) absorption band (discussed further in Sect. 7) of UF_6 reported at 25800 cm^{-1} giving x_{uncorr} = 3.05.

It is well-known[18, 36] that oxo ligands do not have an invariant x_{opt}. Thus, the position of the first electron transfer band of europium(III) varies quite a lot in differing mixed oxides[57, 58] and the major reason seems to be a stabilization due to the Madelung potential dependent on the charges and radii of other cations present[31, 59]. Nevertheless, a large part of the optical electronegativities evaluated for oxide are clustered close to 3.1 and well below 3.5 for water coordinated to an oxidizing central atom. The Stokes threshold of UO_2^{+2} gives x_{uncorr} = 2.4 in the former case, quite compatible with UCl_6 but hardly with UF_6.

Tetrahedral oxo complexes[18, 60] provide fascinating analogies. The purple permanganate MnO_4^-, yellow chromate CrO_4^{-2} and colourless orthovanadate VO_4^{-3} undoubtedly show electron transfer to empty 3d-like orbitals. The lowest wave-number corresponds to excitation of the three M.O. (having the symmetry type t_1 in the point-group T_d) being almost pure oxygen 2p in L.C.A.O. descriptions. The Stokes threshold of the uranyl ion is situated slightly above the wave-number of the first strong absorption band of permanganate, out of all proportion with the chemically oxidizing character of the two complexes. This is particularly striking, when compared with the much higher wave-numbers of the first electron transfer bands of molybdate and tungstate in the ultra-violet. However, again there is a tendency for oxo ligands to make the central atom look less oxidizing than it would have done under other circumstances, as can also be seen by a comparison of MoO_4^{-2} with MoF_6 (to be further discussed in Sec. 7).

It may also be noted[18] that the orange CrO_3X^- have almost the same band positions for X = OH, F, Cl or Br. The (not too evident) explanation seems to be that the first transition goes from a combination of oxygen 2p orbitals (on the three equivalent ligands) to the 3d-like orbitals, which are not very anti-bonding, though they are *strongly* π-anti-bonding in the unsubstituted chromate. It is excluded that the oxo ligands are *much* less reducing than bromide.

Since the lower-lying 5f orbitals are almost non-bonding, we must look for a different explanation for the fact[55, 61] that strongly reducing ligands in the equatorial plane rarely provide new electron transfer bands below the Stokes threshold of the triatomic uranyl ion. Actually, such new bands can frequently be detected at higher wave-numbers than the Stokes threshold, because they are much higher (and much broader) than the typical weak uranyl bands. Thus, it is possible to define an equatorial optical electronegativity $x_{equ} = 1.8$ for U(VI) from the quite intense (ε above 500, superposed the conventional uranyl band with $\varepsilon \sim 10$) band of $UO_2Br_4^{-2}$ at 29800 cm^{-1} as well as $x_{equ} = 2.1$ for Np(VI) from the similar band of $NpO_2Cl_4^{-2}$ at 27700 cm^{-1} [55]. The electron transfer bands due to reducing ligands in the equatorial plane are not only much stronger than the uranyl bands below 30000 cm^{-1}, but they are broad, and like most absorption bands of transition-group complexes at room temperature, do not show vibrational structure.

3 Spin-Orbit Coupling and Group-Theoretical Arguments

Contrary to the usual situation in hexahalide complexes[36] where it is sufficient, to a very good approximation, to consider one-electron energies, it is absolutely necessary to consider the quantum-numbers (written with capital letters, such as S, Λ, Ω in linear symmetries replacing S, L, J in the spherical symmetry of monatomic entities) of the many-electron wave-function[32, 62] in the case of the uranyl ion. However, it is also necessary to start with the orbitals serving as raw material in L.C.A.O. descriptions, such as the empty shells of the central atom in a picture based on uranium(VI) and oxygen(−II):

U 7pπ	U 7pσ			
U 7sσ				
U 6dδ	U 6dπ	U 6dσ		(3)
U 5fφ	U 5fδ	U 5fπ	U 5fσ	

and the filled shells, in part obtained by forming even (g = gerade) and odd (u = ungerade) linear combinations of orbitals of the two oxygen atoms:

Oxygen 2p:	σ_g σ_u π_g π_u	
U 6pπ	U 6pσ	(4)
Oxygen 2s:	σ_g σ_u	

followed by inner shells (such as U 6s, U 5d, ... and O 1s) of which the energies can be determined from photo-electron spectra[47, 48]. The parity of the uranium orbitals is g for even l-values (s and d) and u for odd l-values (p and f). There is one σ orbital in each shell and two orbitals for each of the positive λ values (= 1, 2, 3 ... for π, δ, φ, ...) not higher than l.

Before making a premature attempt to describe optical transitions in the uranyl ion by the electron jumping from a filled to an empty M.O. it is important to realize the large extent of the first-order relativistic effect "spin-orbit coupling" in the U 5f shell. Kaufman and Radziemski[63] found $^2F_{7/2}$ at 7609 cm^{-1} above the groundstate $^2F_{5/2}$ of gaseous U^{+5} representing the two J-levels belonging to the configuration [86] 5f. The concomitant Landé parameter ζ_{5f} is 2174 cm^{-1}. The corresponding J-level energy difference is 4325 cm^{-1} in gaseous Th^{+3} and ζ_{5f} = 1236 cm^{-1} illustrating the tendency[64] first pointed out by Landé of ζ_{nl} being proportional to $(z + 1)^2$ when varying the ionic charge z in an isoelectronic series. It was previously noted[23] that the structured absorption spectrum of the uranyl ion in the near ultra-violet to a surprising extent repeats the structure in the blue, but at 7900 cm^{-1} higher energy.

A specific reason why quantization according to Ω in the point-group $D_{\infty h}$ of linear symmetry including a centre of inversion (and neglecting the neighbour atoms at long distance in the equatorial plane) is appropriate, is that at least the two 5fφ and the two 5fδ orbitals of Eq. (3) have coinciding energies within an interval much narrower than the ^2F separation 7609 cm^{-1} of U^{+5} [53, 65, 66]. An electron in an orbital with positive λ has two alternative values $\omega = (\lambda \pm 1/2)$ whereas one electron in a σ orbital necessarily has $\omega = 1/2$. Another way of counting the levels is that $^2F_{5/2}$ of spherical symmetry provides Ω = 1/2, 3/2 and 5/2 in linear symmetry, whereas $^2F_{7/2}$ provides the same three values, as well as Ω = 7/2. The non-diagonal elements of spin-orbit coupling between such wave-functions having the same Ω were previously given[23] but Fig. 3 of this reference shows the specific result of equidistant 5fλ one-electron energies, which is known today to be a unreliable hypothesis for the almost coinciding λ = 2 and 3. The analogy to Hund vector-coupling (in monatomic entities) is in linear symmetry

$$Q_1 \oplus Q_2 = (Q_1 + Q_2) \text{ or } |Q_1 - Q_2| \tag{5}$$

with the additional remark[32] that when $Q_1 = Q_2$ is positive the result zero has two alternatives 0^+ and 0^- according to whether the many-electron wave-function is invariant or changes sign by reflection in an arbitrarily selected plane containing the axis of linear symmetry. When $Q_1 = Q_2 = 0^+$, the result is 0^+, which is the neutral element of Eq. (5).

We have to perform the vector-coupling Eq. (5) on a situation where one of the orbitals in Eq. (4) is lacking an electron, and where the U 5f shell contains an electron. If one of the σ orbitals in Eq. (4) contains only one electron, Ω = 1/2. If one set of two π orbitals of Eq. (4) contains three electrons, Ω = 1/2 if one (ω = 1/2) is lacking, and Ω = 3/2 if one (ω = 3/2) is lacking. Hence, we need only the multiplication table for a few combinations of $\Omega \oplus \Omega_2$:

	$\Omega_1 = 1/2$	$\Omega_1 = 3/2$	
Ω_2 = 1/2:	$0^+, 0^-, 1$	1, 2	
3/2:	1, 2	$0^+, 0^-, 3$	(6)
5/2:	2, 3	1, 4	
7/2:	3, 4	2, 5	

Since each positive Ω corresponds to two states (in the sense of mutually orthogonal many-electron wave-functions) the transfer of one electron from *either* (ω = 1/2) *or* (ω = 3/2) among the filled M.O. of Eq. (4) to the empty 5f shell corresponds to 28 states. Hence, the blue and near ultra-violet region containing the weak absorption bands of the uranyl ion may correspond to 28, 56 or 84 excited states, allowing a considerable crowding of the Ω levels (16 for each set of 28 states). This book-accounting is not modified by any extent of mixing of states belonging to different manifolds.

It is rather difficult to have preconceived ideas about the order of M.O. energies from Eq. (4). The closest analogy of a dioxo complex with unusually short internuclear distances may be CO_2 having the ionization energies (in eV) determined by photo-electron spectra[49)]

$$\pi_g: 13.8 \qquad \pi_u: 17.6 \qquad \sigma_u: 18.1 \qquad \sigma_g: 19.4 \tag{7}$$

However, the analogy is imperfect, in sofar the empty orbitals of carbon(IV) playing the rôle of Eq. (3) are only C 2s and C 2p with the result that σ_g, σ_u and π_u are expected to dominate the chemical bonding, leaving π_g as non-bonding lone-pairs (cylindrically disposed around the linear axis, on the two oxygen atoms) having the lowest I in Eq. (7). This is not expected to happen in UO_2^{2+} because π_g may participate in bonding with U 6d. Seen from the general view of M.O. involved in electron transfer spectra[32, 36)] the most likely candidate for the lowest I is π_u since U 5f and U 7p are expected to be much less involved in π-bonding than U 6d.

The oscillator strength P (called f by many authors) over the whole first region (between 20000 and 26600 cm^{-1}) of uranyl ions in dilute perchloric acid[23, 24)] is only $9.8 \cdot 10^{-5}$, much lower than the values 0.01 to 0.1 typical[35, 37)] for parity-allowed transitions from odd M.O. in d-group hexahalides. Even the parity-forbidden $(\pi)t_{1g} \rightarrow (5d)t_{2g}$ transition in $IrCl_6^{-2}$ at 17000 cm^{-1} has $P = 0.003$. Further on, we gave the arguments above why at least ten Ω levels are expected to contribute to the blue region of the uranyl ion. An upper limit of P to the lowest-lying Ω level (in the following called the first excited state; or the fluorescent state) includes the two first[24)] "vibrational components" and a part of the third, and is $5 \cdot 10^{-6}$.

The exceptionally low integrated oscillator strength in the region 20000 to 30000 cm^{-1} makes it beyond dispute that only odd M.O. π_u and/or σ_u are excited. It is the general expectation that σ_g (being σ-bonding to both U 6d and U 7s) needs higher excitation energy than π_g (being only π-bonding to U 6d) but the ultra-violet spectrum[22)] only shows a broad, somewhat irregular, increase of very strong absorption up to a maximum close to 50000 cm^{-1} (the beginning of the region needing vacuo studies). It is not likely that electron transfer to U 6d contributes to this strong feature, in view of the positions[55)] of $5f^2 \rightarrow 5f\,6d$ transitions in uranium(IV) hexahalide complexes, and the band[67)] of UF_6^- at 36000 cm^{-1} being the lowest situated, conceivable $5f \rightarrow 6d$ transition. At present, there are two major opinions on the origin of the first absorption bands (in the blue region) of the uranyl ion: that they are $\pi_u \rightarrow 5f$ in analogy to a large amount of evidence obtained from other absorption spectra; or that (at least the lowest excitation energies) are $\sigma_u \rightarrow 5f$ based on arguments by Denning and collaborators, to be discussed below.

Unfortunately, calculations of M.O. energies do not seem able to decide this question, much in the same way[18)] as tetroxo complexes of the permanganate type have a notorious story of calculations yielding nearly all the feasible permutations of one-electron energies. We are looking for differences of excitation energies in the range 1 to 2 eV, and they may already differ from the order of ionization energies by effects related to Eq. (2). Further on, all recent authors agree that relativistic effects are so important that ω rather than λ eigen-functions have to be evaluated. Relativistic modifications of the Xα technique (developed by Johnson and Slater) have been the most popular. The two first columns give the Xα results by Yang, Johnson and Horsley[68] for isolated U^{+2} and oxygen atoms, as well as the uranyl ion, using a one-electron Dirac equation with the muffin-tin potential (U-O distance 1.73 Å) and overlapping spheres (where the radii, of course, introduce a new degree of freedom for the parameters). In the second column are given the ω-values preceded by a parenthesis with the predominant uranium shell or M.O. symmetry (from which the odd or even parity can be deduced). The same quantum numbers are used for the third and fourth column giving the discrete variational Dirac-Slater results by Walch and Ellis[79] for UO_2^{+2} and $UO_2X_6^{-4}$ containing six X^- in a regular hexagon in the equatorial plane with U-X distance 2.43 Å to show the effect of equatorial ligands. 5 fφ is destabilized, pulling the 5 f (ω = 3/2) down to be the lowest empty orbital, and otherwise, only the order of $(\sigma_g)1/2$ and (U 6 p)1/2 are interchanged. Walch and Ellis[79] point out that the zero-point of energy has no physical significance; the two columns are made comparable by choosing 50.9 eV for the lowest orbital. The fifth column gives relativistic self-consistent field results by Boring and Wood[70] not indicating explicitly the spin-orbit coupling separations.

Besides the problem of major importance for us, regarding the relative order of π_u and σ_u there is a more general difficulty for Xα and related techniques applied on polyatomic cations. Four among the seven 5 f-like orbitals have one-electron energies evaluated by Yang, Johnson and Horsley[68] only 0.23 to 1.43 eV above the highest filled M.O. (by the way predicted to be π_g). This discrepancy from the Stokes threshold 2.55 eV is readily explained by Eq. (2), the electron affinity of the 5 f orbitals being far lower than the calculated ionization energy. However, both the U 7 s and U 6 d orbitals are calculated to have ionization energies close to 17 eV, approaching the behaviour of U^{+2}. It can be argued that the distribution in the interval 23 to 27 eV in Table 1 are mainly expected to indicate the relative one-electron energies, but I = 24.5 eV for the highest filled M.O. is out of all proportion while I between 10 and 12 eV known for solid oxides[47] and illustrates the dramatic effect of neglecting a background potential representing the Madelung potential due to adjacent negative charges in an uncharged solid.

Veal, Lam, Carnall and Hoekstra[71] measured photo-electron spectra of several solid uranyl compounds, finding indeed the lowest I values (of essentially O 2 p-like orbitals) close to 11 eV (after a reasonable correction for sample charging during the X-ray bombardment in the photo-electron spectrometer). These authors bring up a very interesting problem, the position of the U 6 p-like orbitals. The calculated values[68] 46.6 eV for U 6 $p_{1/2}$ and 37.0 eV for U 6 $p_{3/2}$ are increased by the similarity to U^{+2} discussed above. It was previously discussed[72] that metallic thorium has Th 6 $p_{3/2}$ 17.3 eV below the Fermi level. The position of U 6 p has not been determined very accurately, as far as we know, but an extrapolation from Z = 90 to 92 and from the metal to the uranyl salts would suggest I (U 6 $p_{3/2}$) close to 25 eV. There is no doubt that the spin-orbit coupling is comparable to the calculated value[68] 9.6 eV, it is already almost 4 eV in the radon atom. A position of U 6 p between O 2 s and O 2 p, as suggested in Eq. (4), opens the interesting possibility of strong participation of U 6 p in the filled M.O. in the region 20 to 35 eV. Indeed, four signals are reported[71] in this region of UO_2CO_3, close to 34, 30, 25 and 20 eV (with charge compensation by comparison with other compounds[47, 48]) and with the valence shell O 2 p at 11 eV) and ascribed to U 6 $p_{1/2}$, O 2 s, and U 6 $p_{3/2}$ with ω = 1/2 and 3/2, respectively.

Table 1. M.O. energies (all in eV = 8065.48 cm^{-1}) calculated for the uranyl ion, and in the first column, monatomic U^{+2} and oxygen atoms. The half-numbered quantum number is ω of linear symmetries. The two last columns give M.O. energies, neglecting effects of spin-orbital coupling, and hence characterized by λ and parity

U^{+2},O^0(Ref. 68)		UO_2^{+2}(Ref. 68)		UO_2^{+2} (Ref. 69)	$UO_2X_6^{-4}$ (Ref. 69)		non-relativistic $X\alpha$ (Ref. 70)	relativistic $X\alpha$ (Ref. 70)
O 2$p_{3/2}$	11.37	(U 5f) 3/2	21.09	–	–	π_u	19.43	16.36
O 2$p_{1/2}$	11.40	(U 5f) 1/2	21.25	18.6	19.5	(U 5f)δ_u	21.74	19.09
U 7s	17.67	(U 5f) 7/2	23.13	–	20.7	(U 5f)φ_u	21.74	19.09
U 6$d_{3/2}$	16.98	(U 5f) 5/2	23.14	21.2	–	σ_u	24.02	23.10
U 5$f_{5/2}$	20.84	(U 5f) 3/2	23.96	22.0	22.6	π_g	24.29	23.47
O 2s	25.80	(U 5f) 5/2	24.33	22.2	22.2	π_u	24.72	23.58
U 6$p_{3/2}$	36.99	(π_g) 3/2	24.56	24.5	25.7	σ_g	25.31	24.94
U 6$p_{1/2}$	46.58	(π_g) 1/2	24.61	24.6	25.9			
U 6s	65.39	(π_u) 1/2	25.51	25.4	26.4			
		(π_u) 3/2	25.65	25.2	26.0			
		(σ_g) 1/2	26.28	23.5	25.6			
		(σ_u) 1/2	26.46	*22.5*	*23.3*			
		(σ_u) 1/2	33.96	31.6	32.8			
		(U 6p) 3/2	38.12	38.6	37.4			
		(σ_g) 1/2	39.85	40.0	43.3			
		(U 6p) 1/2	43.47	43.8	40.5			
		(U 6p) 1/2	50.72	50.9	50.9			
		(U 6s) 1/2	72.52	–	–			

We do not deny that 6p orbitals can be chemically active, e.g if strong oxidants[17] are able to increase the oxidation state of francium(I) or radium(II). However, thorium(IV) aqua ions are as devoid of absorption bands below 48 000 cm^{-1} (6 eV) as barium(II) or lanthanum(III) aqua ions, and we have our doubts whether 6p orbitals could have strong effects in the uranyl ion, though a conceivable counter-argument is the very short U-O distances enhancing the overlap integrals. The major issue would be whether the highest filled M.O. could be σ_u pushed "from below" by 6pσ. In a few cases, photo-electron spectra of gaseous molecules[49, 50] show pushing from below. Thus, the lowest (vertical) I = 15.6 eV of N_2 is due to σ_g followed by I = 16.8 eV due to π_u in spite of the former orbital representing the σ-bonding combination of N 2p, and the two latter M.O. the π-bonding combination of N 2p. This anomaly is removed in O_2; the groundstate of O_2^+ contains one (anti-bonding) π_g and four π_u electrons, whereas the excited states with one π_g, four π_u and only one σ_g electron occur at about 2 eV higher energy than states with three π_u and two σ_g. The pushing upward in N_2 is done by the σ-bonding combination of N 2s orbitals having the same symmetry type σ_g[72] much like the formation of a non-bonding lone-pair[32] in gaseous BH from B 2pσ and B 2s on a L.C.A.O. picture. It is worthwhile to note that nearly all strong effects of "pushing up" are *intra*-atomic, allowing mixture of differing *l*-values. It may also be noted that the N-N internuclear distance is only 1.098 Å, to be compared with C-O 1.163 Å in CO_2 and U-O ~ 1.75 Å in UO_2^{+2}. If σ_u is indeed the M.O. easiest to excite (or to ionize) it would still be surprising if U 6p having a larger I than Zn 3d and Hg 5d in typical zinc(II) and mercury(II) compounds[72] (where the Xnp in dihalides conserve the normal sequence Eq. (7) of energy $\pi_g > \pi_u > \sigma_u > \sigma_g$) should be able to put σ_u above π_u.

Once it was established that several electronic states (Ω levels) are expected[53] slightly above the Stokes threshold of the uranyl ion, it was extremely useful that Denning, Snellgrove and Woodwark[73] studied $Cs_2UO_2Cl_4$ very carefully at liquid helium temperature, and in particular used ^{18}O substitution in order to analyze the vibrational structure and to establish the definite electronic origins. The complex $UO_2Cl_4^{-2}$ exemplifies the tetragonal point-group D_{4h} and has been studied in several other salts, such as formed by the larger cation $N(CH_3)_4^+$ [74]. Flint and Tanner[75, 76] studied the luminescence of $UO_2Cl_4^{-2}$ incorporated in the cubic crystal Cs_2SnCl_6. Though crystallographic studies cannot provide information[6] about the local internuclear distances around such a substituted ion, it is likely that the adaptation to the cubic crystal with Sn-Cl = 2.43 Å modifies to a certain extent U-O = 1.81 and U-Cl = 2.62 Å prevailing in $Cs_2UO_2Cl_4$[76].

The persistence of the uranyl characteristics can be seen from the fact that the Stokes threshold 20600 cm^{-1} only shifts to 20096 cm^{-1} in $Cs_2UO_2Cl_4$[73], 20021 cm^{-1} in $[N(CH_3)_4]_2UO_2Cl_4$[74] and to 19692 cm^{-1} in the substituted Cs_2SnCl_6[76]. Table 2 gives the wave-numbers of electronic origins, and the Ω values proposed by Denning et al.[73] It is interesting to note that the symmetric stretching frequency is 831 cm^{-1} for the groundstate, 714.7 cm^{-1} for the first excited state, and very close to 709 cm^{-1} for the six following electronic states. This close agreement suggests almost parallel potential surfaces. Denning et al.[73] ascribe $\Omega = 1$ to the first excited state, because experiments with polarized light indicate selection rules for magnetic dipolar transitions (which are only possible to $\Omega = 1$ from the groundstate with Ω zero) for the sharp origin (but not for the

Table 2. Electronic origins (all in cm^{-1}) at very low temperatures in crystalline caesium and rubidium uranyl chloride, caesium uranyl nitrate and sodium uranyl acetate. The quantum number Ω characterizing many-electron states in linear chromophores (subject to perceptible relativistic effects) may correspond to two energy levels because of the 4 or 6 ligating atoms in the equatorial plane

Ω:	$Cs_2UO_2Cl_4$ (Ref. 73)	$Rb_2UO_2Cl_4, xH_2O$ (Ref. 244)	$CsUO_2(O_2NO)_3$ (Ref. 77)	$NaUO_2(O_2CCH_3)_3$ (Ref. 77)
1	20095.7	20035	21089	21104
	20097.3	20041	–	21136
2	20406	20408	21694	21915
	21310	21370	–	21926
3	22026	22124	(22300)	21896?
	22076	22143	(23250)?	–
2	22410	22451	23474	(23620)
	22750	22844	–	–
3 (or 5)	26197	26264	26640	23678?
	26247	26281	27480	(27122)?
4	27719	–	–	27795
	27758	–	–	27804

subsequent weak background). Originally, we suggested[53)] $\Omega = 4$ for the first excited state, but[65, 66)] the other result $\Omega = 1$ of the vector coupling (π_u^3 $\Omega = 3/2$) $\oplus$ ($^2F_{5/2}$ $\Omega = 5/2$) of Eq. (6) may have almost the same energy, and perhaps be slightly below or slightly above the other level $\Omega = 4$ according to what uranyl salt is considered. However, partly based on their studies[77)] of other compounds such as $CsUO_2(O_2NO)_3$ and $NaUO_2(O_2CCH_3)_3$, Denning and collaborators believe that $\Omega = 1$ is the result of (σ_u $\Omega = 1/2$) $\oplus$ ($^2F_{5/2}$ $\Omega = 3/2$). One would have expected[65, 66)] the ($\Omega = 5/2$) component to have 1000 to 2000 cm^{-1} lower energy than ($\Omega = 3/2$) of $^2F_{5/2}$ but Denning et al.[77)] assume that the effective one-electron energy of 5 fδ is one or a few thousand cm^{-1} *below* 5 fφ, or that effects of interelectronic repulsion related to Eq. (2) have the same result. At this point, one would have expected the Ω value of the groundstate of the $5f^1$ system NpO_2^{+2} to provide decisive evidence. However, as in so many other areas of this field, the conclusions are rather ambiguous. Paramagnetic resonance[78)] of NpO_2^{+2} syncrystallized in $CsUO_2(O_2NO)_3$ indicates $\Omega = 5/2$ as expected for $^2F_{5/2}$[23, 53, 65, 66)]. However, similar measurements[78)] of $Cs_2UO_2Cl_4$ suggest $\Omega = 1/2$ though one problem may be mixing of Ω values due to the symmetry D_{4h} not being linear. Stafsudd, Leung and Wong[79)] study the absorption spectrum of $NpO_2Cl_4^{-2}$ in $Cs_2UO_2Cl_4$ and conclude that a $^2F_{7/2}$ component occurs at 6880 cm^{-1} but that the four electronic origins observed at 13277, 15426, 17478 and 19358 cm^{-1} are electron transfer bands due to the two oxo ligands. Unfortunately, the $5f^2$ states are expected to be quite crowded, as known from U(IV), Np(V) and Pu(VI).

A related problem is the energy difference between the ($\Omega = 3/2$) and the ($\Omega = 1/2$) components of π_u^3. It may be so small that the excited states overlap. If Denning's hypothesis is correct, Eq. (6) gives exactly the same manifold of Ω values as the ($\Omega = 1/2$) component alone. Returning to the π_u^3 hypothesis, L.C.A.O. formed from pure oxygen 2 p orbitals provide very weak effects of spin-orbit coupling since $\zeta_{2p} = 150$ cm^{-1} only. However, ζ_{7p} known to be 15030 cm^{-1} in gaseous U^{+5}[63)] and ζ_{6p} calculated[68)] to be close to 50000 cm^{-1} suffice that a separation 1000 cm^{-1} between ($\Omega = 3/2$) and the lower-lying ($\Omega = 1/2$) can be produced by 6 percent U 7 p or 2 percent U 6 p, and in the specific case of linear complexes (as contrasted to tetrahedral complexes) the spin-orbit coupling effects of the different atomic orbitals "pull in the same direction" by having the same sign. We return in Sect. 10 to the detailed discussion of the one-electron energy differences in the 5 f shell in neptunyl ions and related species, and we only mention that 5 fσ (participating in the two components with $\Omega = 1/2$) is some 15000 cm^{-1} and 5 fπ probably some 5000 cm^{-1} above the roughly coinciding 5 fδ and 5 fφ. Anyhow, all recent treatments agree that all the electronic origins situated in the interval between 20000 and 25000 cm^{-1} above the groundstate belong to the four categories, where the hole in the filled M.O. provides Ω_1 and where the 5 f-like electron has $\Omega_2 = 5/2$ or $3/2$:

$$\begin{array}{lll} \Omega_1 = 3/2 & \Omega_2 = 5/2 & \Omega = 1 \text{ and } 4 \\ \Omega_1 = 1/2 & \Omega_2 = 5/2 & \Omega = 2 \text{ and } 3 \\ \Omega_1 = 3/2 & \Omega_2 = 3/2 & \Omega = 0^+, 0^- \text{ and } 3 \\ \Omega_1 = 1/2 & \Omega_2 = 3/2 & \Omega = 1 \text{ and } 2 \end{array} \qquad (8)$$

The hypothesis by Denning, Snellgrove and Woodwark[73)] of σ_u excitation only utilizes the second and the fourth line of Eq. (8), and we already mentioned that their assignment of $\Omega = 1$ to the first excited state then belongs to the fourth line. These authors[73] make a distinction between even and odd Ω values by the relatively strong separation of an even Ω in two electronic origins in the symmetry D_{4h} of the complex $UO_2Cl_4^{-2}$ whereas odd Ω values should not be influenced by this tetragonal symmetry, but only from the weak orthorhombic distortion in the crystal to the local point-group D_{2h} which is expected to have far weaker energetic consequences. As seen in Table 2 two distinct polarization selection rules apply to the components a and b. These sets of electronic origins are separated by 904 and by 340 cm^{-1} in the two cases assigned $\Omega = 2$, and by 39 cm^{-1} in the case identified as $\Omega = 4$, whereas the separations in the two cases of $\Omega = 3$ are only 50 cm^{-1}. The second case of $\Omega = 3$ may rather be[73] $\Omega = 5$ which is a unique situation in Eq. (6) since it can be simultaneously described[53] by the vector coupling (π_u^3 $\Omega = 3/2$) $\oplus$ ($^2F_{7/2}$ $\Omega = 7/2$) and (most unusually) by the quantum numbers $S = 1$ and $\Lambda = 4$. This debatable state (its vibrational frequency is exceptionally high, 724 cm^{-1}) occurs at 6100 cm^{-1} above the first excited state of $Cs_2UO_2Cl_4$ (cf. Table 2) whereas the states identified as $\Omega = 4$ at 7600 cm^{-1} (with vibrational frequency 706 cm^{-1}) are comparable to the second set of excited states clearly starting[23] at 7900 cm^{-1} above the first set of the uranyl ion in aqueous solution. In our opinion, the second set is a closely crowded collection of Ω levels obtained by vector coupling of $\Omega_1 = 3/2$ and 1/2 with the almost coinciding $\Omega_2 = 7/2$ and 5/2 of $^2F_{7/2}$. To a very good approximation the second set is expected to have the same energy above the Stokes threshold as $^2F_{7/2}$ above $^2F_{5/2}$.

The fundamental problem behind the Ω identifications in Table 2 in whether *all* the electronic origins have been detected. This is not only a question of relative oscillator strength (i.e. area of the absorption bands) but also on the sharpness of the electronic origins. It is true that the assignments of the origins between 20096 and 22750 cm^{-1} have exactly the Ω values 1, 2, 3 and 2 obtained by combining the second and the fourth lines of Eq. (8). However, how certain is it that some electronic transitions are not hiding away in the background (or accidentally coinciding with a vibrational component) and how certain are the Ω assignments? We are surprised that the local orthorhombic perturbations have as large effects, when compared to the splitting[73] below 2 cm^{-1} of the first excited state $\Omega = 1$. Denning et al.[77] also ascribe $\Omega = 1$ to the first excited state of the trigonal complexes $UO_2(O_2NO)_3^-$ and $UO_2(O_2CCH_3)_3^-$. The sodium salt of the latter anion show several multiple origins, ascribed to factor group splitting, but we still have our doubts whether a few of these features do not have other Ω values. It is slightly worrying that the distance from $\Omega = 1$ to the first level with $\Omega = 2$ increases from 605 cm^{-1} in the nitrate to 801 cm^{-1} in the acetate, though this order of magnitude would fit the spin-orbit splitting expected in π_u. A constructive argument would be detecting the elusive $\Omega = 4$.

The general influence by the equatorial ligands has also been investigated by Görller-Walrand et al.[80–82]. In particular, three bidentate ligands (such as nitrate or acetate) provide the holohedrized symmetry[32] D_{6h} (however much the point-group is D_{3h}) and only Ω values which are multiples of 3 should separate the energy of their two states. This should be an excellent opportunity to recognize the Ω values of Eq. (8) but the only fairly convincing case of a split pair of $\Omega = 3$ states in Table 2 is 26640 and 27480 cm^{-1} in the double nitrate, situated below the expected $^2F_{7/2}$ limit. Denning et al.[77] mention the possibility that this, second, instance of $\Omega = 3$ may be due to π_u excitation starting at slightly higher wave-numbers than σ_u excitation, allowing altogether for 48 Ω levels formed by the 84 states representing twice the left column, and once the right column, of Eq. (6). Recent work[247] involves the effects of ^{15}N substitution on $Cs(UO_2)(O_2NO)_3$.

Professor Colin D. Flint was so kind as to provide us with several helpful preprints, relating spectra[83] of the pentagonal $UO_2F_5^{-3}$ (the two lowest origins occur at 19982 and 19987 cm^{-1} in the potassium and at 20022 and 20099 cm^{-1} in the ammonium salt), of hydrated salts[244] containing $UO_2Cl_4^{-2}$, at 85 K τ is found as high as 3.6 ms in $[N(CH_3)_4]_2UO_2Cl_4$ corresponding to a higher limit of $P = 2 \cdot 10^{-6}$ in Eq. (9) below, and in spite of the reducing character of bromide, $Cs_2UO_2Br_4$ is found[246] to luminesce from origins at 19665 and 19674 cm^{-1}. Most, but not all, of these observations are compatible with Dennings hypothesis.

Fluorescent organic molecules have very distinct triplet ($S = 1$) *or* singlet (S zero) character of their excited states. For a long time, the same question has been asked[2, 22] about the first excited state of the uranyl ion. It is conceivable that some electron transfer states, e.g. of orthovanadate[60, 84] are quite well-defined triplet states. However, with exception of the unique case $\Omega = 5$, we emphasized[53] that the coupling scheme Eq. (6) would leave 20 to 80 percent singlet character (and the complement triplet character) in the eigen-values, and that Russell-Saunders coupling is broken down even more completely, because there are five alternative Λ values (from 0 to 4) to be mixed in the eigenstates. Hence, it has no sense to look for Ω-components of $^3\Pi$ or $^3\Delta$ or whatever terms have been proposed in literature. A particular aspect of Eq. (6) is that *if* $5f\varphi$ and $5f\delta$ have identical one-electron energies, the lowest eigen-value $\Omega_2 = 5/2$ has exactly the same distance from the higher $\Omega_2 = 5/2$ (coinciding with $\Omega_2 = 7/2$) as $^2F_{5/2}$ up to $^2F_{7/2}$ in spherical symmetry[65, 66]. The uranyl ion exemplifies a rather unique coupling scheme[62] where Ω_2 corresponds to $(SLJ\Omega_2)$ parentage, whereas Ω_1 is much more similar to the usual behaviour of linear molecules.

4 Hydrogen Atom Abstraction and Complexes of the Excited State with Anions

We may make two relatively trivial statements: that an excited state of a species in condensed matter (liquids, and amorphous, vitreous and crystalline solids) luminesces, if it does not decay by other mechanisms (such as multi-phonon desexcitation[33] producing heat, or by photochemical reactions); and that a reasonably long-lived excited species (10^{-7} s is more than sufficient) has thermalized to the Boltzmann distribution of its own vibrational states, with the result that it is, at the same time, *much more oxidizing* and *much more reducing* than the groundstate[85–87]. The quantitative form of the latter statement is that the standard oxidation potential has changed from E_0 to $(E_0 + h\nu)$ as an oxidant, and the species as a reducing agent has changed from E_0' to $(E_0' - h\nu)$, where $h\nu$ is the photon energy (1 eV = 8065.48 cm^{-1}) at the Stokes threshold. Even when taking the kinetic barriers of oxygen and hydrogen "overvoltage" into account, it is very difficult to prevent an oxidizing species with the corresponding E_0 above + 1.5 V from evolving O_2 in aqueous solution, or a reducing species with E_0 below − 0.5 V from evolving H_2. Hence, visible light with photon energies between 1.6 and 3.1 eV is expected to bring either the reducing, or the oxidizing character (or both) of most luminescent species outside this range compatible with aqueous solutions. Seen from this point of view, the technologically desirable decomposition of water by solar light to H_2 and O_2 should have been much easier than it actually is[88–91]. A chemist may want an example of chemical substitution enhancing simultaneously the oxidizing and the reducing character of the groundstate (without illumination). A simple case is H_2O_2 being, at the same time, much more oxidizing than H_2O and more reducing, e.g. toward cerium(IV) and permanganate in acidic solution.

Though it is difficult to study UO_2^+ outside a narrow pH interval around 2, since it disproportionates rapidly to U(VI) and U(IV) outside this interval (to UO_2^{+2} and uranium(IV) aqua ions at lower pH, and to almost insoluble hydroxides and salts of the type $Na_2U_2O_7$ at higher pH), E_0 for the oxidation of UO_2^+ to UO_2^{+2} has been determined to $+0.06$ V relative to the standard hydrogen electrode[92]. With the arguments given above, E_0 of the excited uranyl state is $+2.6$ V, almost as high as for fluoride anions to F_2[1, 65, 93]. Strictly speaking, this conclusion involves an auxiliary argument, that the removal of the 5f electron from UO_2^+ to form the groundstate of UO_2^{+2} involves roughly the same amount of energy related to rearrangement of internuclear distances as letting the 5f electron present in the excited state of UO_2^{+2} jump back to the hole in the M.O. filled in the groundstate. However, this contribution should be considerably smaller than the average Stokes shift by uranyl fluorescence, and hence smaller than 0.1 eV.

Once it is realized that the fluorescent state of the uranyl ion is almost as oxidizing as free fluorine, it is not surprising that it performs rather violent photochemical reactions. The rather complicated and unexpected reactions with organic molecules[2, 22] usually can be rationalized as organic rearrangements subsequent to transfer of a hydrogen atom to the excited uranyl ion. In view of the strong affinity to certain anions discussed below, and the presence of U(IV) in the final products of the photochemical reaction, one might imagine transfer of a hydride ligand. However, it seems almost certain[94, 95] that the primary product is U(V), and most probably UO_2^+ which then disproportionates in the presence of water. It is noted that direct transfer of an electron is among the conceivable origins of UO_2^+ and as discussed in Sect. 5, such reactions seem to occur with some inorganic cations. The fulgurant attack on organic molecules is reminiscent of radiochemistry, though the uranyl photochemistry is not at all dependent on the very weak radioactivity of ^{238}U. It was previously mentioned[1, 33] that one of us in 1944 exposed mixtures of uranyl nitrate dissolved in 4 molar sulphuric acid, and ethanol, to sunlight with the purpose of precipitating the salt $U(SO_4)_2$, 4 H_2O. The supernatant solution smelled exactly like old Madeira wine. It would be interesting to study such samples with modern analytical techniques, such as gas chromatography and mass spectrometry. It may be that the hydrogen abstraction by the excited uranyl ion provides complicated polycondensed molecules (much like the slow air oxidation of distilled wine to cognac or brandy) in addition to monomeric acetaldehyde, and its adduct acetal with an ethanol molecule.

It is not all organic compounds that loose hydrogen atoms to the excited uranyl ion. Thus, the quantum yield of luminescence of many binary and ternary acetates is quite high. The ternary acetates probably all involve the anion $UO_2(O_2CCH_3)_3^-$ containing six oxygen atoms in the equatorial plane, coordinated from three bidentate ligands. A well-known case is $NaZn(UO_2)_3(CH_3CO_2)_9$, 6 H_2O used for gravimetric determination of sodium, the molar weight being 67 times the atomic weight of Na. The careful studies of the photodecomposition of the hydrated formiate $UO_2(HCO_2)_2$, H_2O demonstrate the rôle played by the water molecule[94, 96]. On the other hand, it is not known whether the uranyl oxalate complex used as actinometric solution[2, 22] contains constitutional water. It is remembered that many other oxalate complexes decompose in ultra-violet radiation. As an example of electron (rather than hydrogen atom) transfer may be mentioned[254] the triplet adduct between UO_2^+ and NO_3 formed at 4 K by u.v. irradiation of crystalline $UO_2(O_2NO)_2\,(OP(OC_2H_5)_3)_2$. The thermal barrier against electron back transfer is about 100 cm^{-1} and the adduct decays rapidly above 10 K.

It is evident that kinetic factors decide whether a species as oxidizing as the excited uranyl ion has the time to react with surrounding species during its life-time, and also decide the relative probabilities of several alternative photochemical reactions. The mean life-time τ of the uranyl ion in dilute aqueous non-complexing acids is remarkably short, close to 1.2 μs ($1.2 \cdot 10^{-6}$ s) and corresponds to a quantum yield of fluorescence well below 1 percent. The fact[2] that the life-time is only about twice as long in deuterium oxide D_2O suggests that the non-radiative desexcitation is not a straightforward multi-phonon process involving O-H stretching frequencies, as is the case[97, 98] for the low-yield line fluorescence of $4f^5$ samarium(III), $4f^6$ europium(III) and $4f^8$ terbium(III) aqua ions which is enhanced by the factors 26, 54 and 12, respectively, when going from H_2O to D_2O. This argument does not apply to the ultra-violet luminescence at 32200 cm^{-1} of $4f^7$ gadolinium(III) aqua ions, which does not suffer perceptible competition with multi-phenon desexcitation[99]. The influence of the adjacent water molecules on the excited uranyl ion seems rather to be due to abstraction of a hydrogen atom. The remaining hydroxyl radical OH recombines with the uranium(V) intermediate, producing no over-all photochemical reaction. Marcantonatos[100–102] has analyzed the inner mechanism in great detail, and argues that the monomeric product $^*UO_2H^{+2}$ (we write the excited uranium atoms with an asterisk) participates in a Brønsted acid-base equilibrium with H_3O^+ and an excited $^*UO_2^+$ only decaying by non-radiative processes to the groundstate UO_2^+. The $^*UO_2H^{+2}$ may either produce luminescence with a low yield, giving an emission spectrum rather similar to that of conventional $^*UO_2^{+2}$ (we return below in Sect. 5 to the question of the exciplex formed between groundstate UO_2^{+2} and $^*UO_2H^{+2}$) or it may perhaps show no detectable luminescence.

It may be important for the elucidation of these complicated questions that Moriyasu, Yokoyama and Ikeda[103] pointed out that τ for exchange of the equatorial water molecules in the groundstate UO_2^{+2} at 25 °C is 0.8 μs, slightly shorter than τ = 1.2 μs for luminescence of uranyl aqua ions, and they argue that the exchange of water should be more rapid in $^*UO_2^{+2}$ having uranium(V) characteristics due to the screening of the 5f electron. Unfortunately, the rates of water exchange in aqua ions seem difficult to predict, e.g. in lanthanides[5, 6]. Anyhow, the luminescence τ = 265 μs for crystalline UO_2SO_4, 3 H_2O at 25 °C and τ = 425 μs for the corresponding UO_2SO_4, 3 D_2O[104] clearly show that the aqueous solution has decreased τ by a factor close to 200, which does not preclude the presence of equatorial water molecules in the crystal. Anhydrous crystals[104, 105] can show τ as high as 1400 μs.

If the excited level contains e_2 states $h\nu$ above the groundstate (representing e_1 mutually orthogonal states), Einstein's formula from 1917[33] relates the radiative life-time τ_{rad} to the oscillator strength P of absorption:

$$\tau_{rad} = \frac{e_2}{e_1} \cdot \frac{2.3 \cdot 10^{-8}\,s}{P(h\nu/eV)^2} \tag{9}$$

We mentioned above that if the structure above the Stokes threshold is analyzed, P of the *first* excited level cannot be higher than $5 \cdot 10^{-6}$. Combining this value with $e_2 = 2$ (one positive Ω value), $e_1 = 1$ and $h\nu = 2.6$ eV, τ_{rad} is calculated to be $1.4 \cdot 10^{-3}$ s, exactly the higher limit known for anhydrous crystals[104, 105]. Though the number of vibronic states may produce problems for Eq. (9) there is no doubt that $\tau = 425$ µs reported[104] for UO_2SO_4, 3 D_2O already suggests quantum yields η close to 0.3 since it is generally valid[33] that $\tau_{obs} = \eta\,\tau_{rad}$. This is one reason why independent determination of quantum yields are highly useful in problems of this kind, where the value of (e_2/e_1) may not be evident, as it is in monatomic entities. If the half-life $t_{1/2} = \tau \ln 2$ is wanted, the time constant of Eq. (9) is $1.6 \cdot 10^{-8}$ s.

It is known since long time[22] that reducing (and polarizable) anions such as Br^-, I^- and NCS^- quench the luminescence of uranyl aqua ions even much more efficiently then water itself. These studies were continued by Moriyasu et al.[103]. One might have expected that the quenching is due to very short-lived collision complexes, since the groundstate uranyl ion is known to form quite strong complexes with fluoride and with oxygen-containing (especially multidentate) ligands, but at most outer-sphere ion-pairs with bromide and iodide. As discussed by Ahrland[106] the former type of uranyl complexes have high formation constants corresponding to negative ΔG values, but usually positive ΔH indicating endothermic absorption of heat. However, it is clear today that an inorganic species living for so long time as 10^{-6} to 10^{-4} s may have its own chemistry, in this case determined by its oxidizing character being more pronounced than of any high oxidation state of a noble metal. Preliminary measurements by Professor Marcantonatos suggest a large formation constant of Br^- with $*UO_2^{+2}$ and it has been shown that the excited state not only forms rather strong complexes with nitrate[107] but also with perchlorate[108] in spite of the opinion prevailing among complex chemists that perchlorate does not form inner-sphere complexes in aqueous solution[3]. It is interesting to compare[109] with palladium(II) forming (at pH = 0) a complex with one NO_3^- with the formation constant K_1 close to 1.2 l/mol, though one has to remember[6] that such small "constants" are rather dependent on the concentrations of other ions present. On the other hand, the aqua ion $Pd(OH_2)_4^{+2}$ does not seem[109] to form comparable complexes with perchlorate nor with *p*-toluenesulphonate, and at least, the absorption spectrum is not influenced to any significant extent by varying concentrations of these two anions. The complex[107] $*UO_2NO_3^+$ has $K_1 = 1.77$ l/mol at 1 °C but only 0.36 at 25 °C. K_1 for the groundstate $UO_2NO_3^+$ was found be 0.25 l/mol at 25 °C in agreement with literature, but the formation of this complex is endothermic, and K_1 increases with temperature. The average life-time τ for $*UO_2NO_3^+$ is about half the values for $*UO_2^{+2}$ indicated[107] to vary between 8 µs at 1 °C and 1.3 µs at 25 °C. It is argued that $*UO_2NO_3^+$ exchanges its nitrate ligand much more rapid than its decay. It may be noted that the absorption spectrum of $UO_2NO_3^+$ is about twice as intense in the blue as uncomplexed UO_2^{+2}. At pH = 2, Moriyasu et al.[103] find τ increasing from 1.2 µs for 0.1 molar ClO_4^- to 1.6 µs for 1 molar solutions. In the latter case[108] the absorption spectrum is hardly modified and the steady-state luminescence grows proportionally more intense at all wave-lengths. It seems that a weak complex $*UO_2ClO_4^+$ is formed with $K_1 = 0.32$ l/mol, and having the enhanced $\tau = 43$ µs, at 25 °C.

An even more dramatic effect is observed[103] for $*UO_2F^+$ having $\tau = 77$ µs. Deschaux and Marcantonatos[110] propose another interpretation involving an exciplex, and argue that the situation is the opposite of the perchlorate case, K_1 of $*UO_2F^+$ being too small to be evaluated, whereas it is as high[106] as 34300 for the groundstate UO_2F^+ again suggesting behaviour like a noble metal of the excited uranyl ion, comparing to its conventional chemistry. Actually, the study of fluoride complexes[106] is difficult, because $K_2 \sim 2700$ and $K_3 \sim 280$ for formation of UO_2F_2 and $UO_2F_3^-$ are quite large, and because the Brønsted acid having pK around 3 (which does not seem to be diatomic HF, but rather[111] the strongly bound ion-pair $OH_3^+F^-$) and the strongly hydrogen-bonded FHF^- remove F^- from equilibrium at lower pH. Matsushima et al.[112] also noted much longer τ and higher quantum yield η of luminescence for the uranyl ion dissolved in 0.67 molar phosphoric acid. Moriyasu et al.[103] found $\tau = 178$ µs in 1 molar H_3PO_4 and 11 µs in 1 molar H_2SO_4. If the τ of the uranyl aqua ion is about 1000 times shorter than τ_{rad} because of intramolecular rearrangements involving ephemeral abstraction of hydrogen atoms from the coordinated water molecules[100, 102] it cannot be argued that species such as $*UO_2F^+$, $*UO_2ClO_4^+$ or the complexes of the excited uranyl ion with one or several (more or less protonated) phosphate ions[93] contain much fewer coordinated water molecules. So long at least one constitutional water molecule remains, it cannot be argued that τ of $*UO_2^{+2}$ increases by a factor of 30 to 150 because the hydrogen atoms are no longer available. The explanation must involve kinetic barriers against this hydrogen abstraction from surrounding water molecules in the anion complexes, except $*UO_2(NO_3)^+$ and may be related to our ignorance of the detailed hydrogen bonding around aqua ions. Professor Marcantonatos was so kind as to draw our attention to the possibility[102] that the ephemeral hydrogen atom abstraction does not occur in practice from a large number of water molecules loosely coordinated in the equatorial plane, but rather from a single molecule very close to one of the two oxygen atoms of $*UO_2^{+2}$ lacking the electronic density that has been transferred to the 5f shell.

It is a well-known characteristic of the uranyl ion (together with thorium and beryllium) in qualitative analysis that it is soluble in ammonium carbonate, though it is not soluble in aqueous ammonia (such as nickel, copper, zinc, ...). In strong HCO_3^- solutions (in order to diminish the formation of hydroxo complexes) it seems that $UO_2(O_2CO)_3^{-4}$ occurs, analogous to the crystalline double nitrates, but otherwise, complicated mixed hydroxo-carbonate complexes are formed[113]. By the way, uranium(V) also forms very strong carbonate complexes[114] what is also true for NpO_2^+. The oceans contain close to 10^{-8} molar uranyl compounds. In view of pH slightly above 7 and the high concentration of $HOCO_2^-$ there is little doubt that the dissolved species are uranyl carbonate complexes. By the way, the concentrations of otherwise common elements such as aluminium and iron(III) in true (not colloidal) solution are lower, because of the very low solubility products of the hydroxides. In the water of certain lakes and rivers, the concentrations of uranyl carbonate are considerably higher than in sea water.

Test tube experiments show much stronger uranyl fluorescence in $HOCO_2^-$ solutions (much like in phosphoric acid) than of the aqua ion. It was discussed[1] whether the 10^{-8} molar uranyl carbonate in the uppermost 50 m of the ocean participates in the photochemical oxidation of (otherwise not biodegradable) organic compounds arriving from the great rivers (such as Amazonas and Zaïre) which would otherwise accumulate in the oceans. At standard conditions AM1 (i.e. the sun standing in zenith) the solar radiation[25] arriving at the Earth's surface amounts to about 1 kW/m^2. If scattered light is included, about 200 W/m^2 has photon energies above 20000 cm^{-1}. With an average photon energy 22500 cm^{-1} in this region, the number of blue photons per m^2 is $4 \cdot 10^{20}$ per second, or 2.4 einstein/h where 1 einstein is a mole ($6 \cdot 10^{23}$) of photons, like 1 faraday (96485 C) is the charge of a mole of protons. As long the uranyl complex does not produce a large part of the total light absorption in the sea water, and if $\varepsilon = 10$, each uranyl ion is excited, on the average, every 10 min, corresponding to an overall photochemical turn-over rate of 0.0005 mol/h in a column 1 m^2 in area perpendicular on the solar radiation, and 10 m deep, containing 10^{-8} molar uranyl carbonate complexes, and hence absorbing 10^{-4} of the incoming blue light.

It is difficult to tell whether this turn-over rate is multiplied exactly by the quantum yield of fluorescence (the non-radiative processes not contributing to the probability of photochemical reactions) which may still be several percent in the carbonate solution. For instance, the non-biodegradable compounds may form strong complexes (via carboxylate or phenolic hydroxo groups) with the uranyl groundstate, but otherwise, one would expect most of the photo-chemistry to follow molecular collisions with the excited uranyl carbonate complex. Professor Burrows was so kind as to draw our attention to the competing quenching by the $8 \cdot 10^{-4}$ molar Br^- present in sea water, which may take over a substantial part of the excitation energy[115]. On the other hand, the 0.5 molar Cl^- and $4 \cdot 10^{-7}$ molar I^- should be less of a problem, chloride because of the much less efficient quenching, and iodide because of the concentration 2000 times lower than of bromide.

5 Exciplexes and Complexes of the Excited State with Cations

The formation of an *exciplex* is due to the affinity of the relatively long-lived excited state of a molecule to a groundstate molecule. It is possible to have comparable hetero-exciplexes between two differing species, also including deuterium-substituted aromatic hydrocarbons (where the deuteriation is known to prolongate the life-time of "phosphorescent" triplet states by a large factor). The classical case is pyrene, where the monomeric molecule emits violet fluorescence from a singlet state, showing conspicuous vibrational structure, whereas the exciplex emits a broad band in the blue (at lower energy). It is possible to use arguments based on the mass-action law. The formation of the excimer is enhanced by higher concenctration in a given solvent, and by higher temperature. In particular, cooling toward − 100 °C provides nearly pure monomer fluorescence[116]. Excimers exist not only in condensed matter, but are quite frequent in the gaseous state, in particular in electric discharges.

The straightforward exciplex between $^*UO_2^{+2}$ and UO_2^{+2} has not been reported, to our knowledge. However, Marcantonatos[117] found evidence for an exciplex between $^*UO_2H^{+2}$ (the primary product of hydrogen atom abstraction from a coordinated water molecule) and UO_2^{+2}. Unless $^*UO_2HO_2U^{+4}$ has a symmetric hydrogen bond like FHF^- (what seems unlikely) it is strictly speaking a hetero-exciplex, but it is the first well-characterized, purely inorganic exciplex. The emission of the exciplex at pH = 1.94 (dilute nitric acid) is shifted to slightly lower photon energies, and the vibrational frequency in the approximately equidistant progression is about 120 cm^{-1} lower than in $^*UO_2^{+2}$ suggesting a weakening of the U–O bond in the exciplex. Both of these tendencies are compatible with a larger Stokes shift. Further studies[101, 102] of this exciplex by time-resolved emission spectra allowed its mean life-time to be measured to be 6 μs, and a time-window at 8 to 10 μs after the primary excitation permitted a much more clear-cut fluorescence spectrum to be evaluated, with rather irregular structure below 18 000 cm^{-1} (550 nm). At millimolar and higher concentrations of the uranyl aqua ion, the hydroxo complexes formed by increasing pH are oligomers. In particular[118] a predominant species is $UO_2(OH)_2UO_2^{+2}$ which is known from X-ray diffraction of buffered solutions and in crystal structures[245] to contain two hydroxide bridges in the equatorial plane. The equilibrium constant $^*\beta_{22} = [(UO_2)_2(OH)_2^{+2}][H^+]^2/[UO_2^{+2}]^2 = 10^{-6}$ mol/l[118] corresponds to 0.1% of the 0.1 molar uranyl solution at pH = 2 existing as this dimer. If its formation and dissociation were very rapid, it could help to understand a strongly decreasing τ as a function of higher pH and of higher total uranyl concentration (as is most likely to be observed at pH well above 2) but the observation of five times longer τ in the exciplex would rather be explained by the dimeric complex dissociating slowly, and trapping an excited state which has given up a large part of its propensity to intramolecular hydrogen atom absorption (such as $^*UO_2F^+$ or the solutions in phosphoric acid). However, the kinetic arguments by Marcantonatos[101, 102, 117] are against this explanation.

In their study[110] of $^*UO_2F^+$, Deschaux and Marcantonatos argued that a comparable exciplex $FO^*UOHOUO^{+3}$ is formed (at very low fluoride concentrations in the 10^{-4} to 10^{-3} molar range) by the interaction between $^*FUO_2H^+$ (having abstracted a hydrogen atom from a water molecule) and UO_2^{+2}. As seen on their Fig. 2, the emission spectrum of the exciplex is almost shifted 200 cm^{-1} toward lower energies.

Burrows et al.[119] studied the quenching of uranyl fluorescence in 0.01 molar HNO_3 by a large number of metal aqua ions. One of the more spectacular results is the formation (with a rather low quantum yield) of manganese(III) aqua ions (with a characteristic absorption band in the green) by flash photolysis in the presence of $Mn(OH_2)_6^{+2}$ showing the transfer of one electron from the latter species. Marcantonatos[93] continued such studies in 0.1 molar HNO_3 containing 0.7 molar additional nitrate, and these authors agree that the quenching varies, for a giving concentration of the quenching cation:

$$Tl(I) \sim Ag(I) > Pb(II) \gg Mn(II) > Cu(II) > \ldots \quad (10)$$

Hg(II) is reported[93] to be comparable to Cu(II), but there is a risk of forming traces of the mercurous ion Hg_2^{+2} situated[119] between Ag(I) and Pb(II) in Eq. (10). Fe(II) is also a very effective quencher[119] but the absorption spectrum of mixtures with UO_2^{+2} is not additive in the ultra-violet[93] much like the influence[120] of SO_2 on $Fe(OH_2)_6^{+2}$ Ce(III) is also quenching, but to a much smaller extent than expected from its chemically reducing character, and the reaction mechanism seems different[93]. Matsushima et al.[112] previously found several other aqua ions to quench to a certain extent. There seem to occur very weak quenching effects[93] of Zn(II), Ga(III), Sr(II), Cd(II) and Ba(II).

The major, if not unique, origin of the quenching described by Eq. (10) is the reducing character of the aqua ion. However, the quenching rate is not determined exclusively by the chemical ionization energy I_{chem}[10, 31, 32] but also by the spatial extension of the orbital containing the electron to be transferred[93]. Thus, the strongly localized 4f electron[33] of cerium(III) is much less efficient for quenching of $^*UO_2^{+2}$ than the extended 6s-like orbital[32, 41] of thallium(I) and lead(II). Seen from the point of view[10, 31] of standard oxidation potentials E_0 many chemists are surprised to see silver(I) and thallium(I) as the two most efficient reducing agents in Eq. (10). However, contrary to most M(II) and M(III), these two M(I) are not strongly coordinated by water molecules, at least not in all directions (this allusion refers to linear complexes such as $Ag(NH_3)_2^+$). Hence, the unsolvated cations provide new absorption bands[121, 122] in salts of oxidizing anions containing a partly filled 5 d-shell, such as Re(IV), Os(IV) and Ir(IV) hexahalide complexes, or an empty 3 d-shell, such as chromate and permanganate. Marcantonatos and Deschaux[123] give the arguments for the strongly bound $O{*}UOAg^{+3}$ to have Ag^+ bound to one of the oxygen atoms. Since each of the oxygen atoms in $*UO_2^{+2}$ carry almost half an electronic charge less than in the groundstate, purely electrostatic bonding counteracts the binding of Ag^+ in the excited state, and it is the covalent characteristics of silver(I) that are involved, much like the strong affinity of Pt(IV) or Au(III) to the heavier halides. Marcantonatos and Deschaux[124, 125] also find that thallium(I) forms not only $O{*}UOTl^{+3}$ but also $TlO{*}UOTl^{+4}$ where each oxygen atom is bound to a cation. The equilibrium constant for binding the first Tl^+ to $*UO_2^{+2}$ seems to be close to 750 l/mol and for the second only of the order of magnitude 1 l/mol. The latter hetero-exciplex is fluorescent to a detectable extent, what is not the case for $O{*}UOTl^{+3}$ (probably dissociating[125]) readily in UO_2^+ and Tl^{+2}).

In kinetic studies, it is generally expected that the reaction rates between cations are influenced by the anions present, and reaction rates between anions by the cations present Deviations from this rule were called "anomalous salt effects" by Brønsted. In this sense, the reactions between $*UO_2^{+2}$ and Ag^+ or Tl^+ correspond to intermediates of anomalous salt effects. However, the reciprocal dielectric constant 0.013 of water is so low that Coulomb interactions by no means are predominant. For our purposes, the complexes formed by the neptunyl(V) NpO_2^+ with other cations are quite relevant. Contrary to the rapidly disproportionating UO_2^+, the hydrated NpO_2^+ can be studied in acidic solution, showing a large number of narrow absorption bands due to the presence of two 5f electrons. Contrary to the original idea of Gruen[126] which is much more appropriate for uranium(IV) complexes[31, 55] it cannot be argued that each absorption band of NpO_2^+ corresponds to a definite J-level[33] of the configuration $5f^2$. Actually, the "ligand field" effects corresponding to σ-anti-bonding of $5f\sigma$ and to π-anti-bonding $5f\pi$ orbitals of Eq. (3) are expected to split each of the thirteen J-levels into $(J + 1)$ different Ω levels (in the prevailing linear symmetry) and differing levels with the same Ω are able to mix their J-characteristics. Unfortunately, the detailed problem of identifying the 52 Ω levels (formed by the 91 states of $5f^2$) has not yet been solved, and for our purposes, it is sufficient to consider the absorption bands as pragmatic entities. An exceptionally intense ($\varepsilon = 387$) and very narrow absorption band of hydrated NpO_2^+ occurs at 10200 cm^{-1}. If trivalent aqua ions, such as aluminium, scandium, iron or indium; or thorium(IV) aqua ions, are added, this band shifts between 10020 and 10110 cm^{-1} and decreases the molar extinction coefficient ε to 228 for Cr(III)[127]. When the reactions of the M(III) are slow, it is possible by ion-exchange resin techniques to isolate solutions of the green $(H_2O)_5CrONpO^{+4}$ [128] and of the even slower reacting rhodium(III) complex $(H_2O)_5RhONpO^{+4}$ [129]. The $3d^3$ absorption bands of the chromium(III) complex leave no doubt that the unidentate ligand $ONpO^+$ has replaced one of the water molecules in the hexaqua ion.

When chromium(II) aqua ions are oxidized by UO_2^{+2} a similar $(H_2O)_5CrOUO^{+4}$ is obtained[130] though it rearranges rather rapidly. It is even possible[131] to detect an adduct between groundstate UO_2^{+2} and UO_2^+ though the formation constant is only 15 l/mol. This adduct is the closest analogy we know to the exciplex $*UO_2HO_2U^{+4}$ though it probably has one oxygen atom of UO_2^+ bound to the equatorial plane of UO_2^{+2}. Because of the low formation constant of the adduct, the exciplex[102] dissociates, for practical purposes, irreversibly to H_3O^+, UO_2^+ and UO_2^{+2}. Unfortunately, the seven Ω levels provided by the 5f electron in the uranium(V) have not been investigated by absorption spectra. The reactions of NpO_2^+ with hydrated cations have been further reviewed[132]. We return in Sect. 10 to the rather conflicting spectroscopic evidence on UO_2^+ aqua and carbonate complexes, and to the adducts $U_2O_4^{+3}$ and $UNpO_4^{+3}$.

6 Absorption Spectrum and Other Properties of the Excited Uranyl Ion

It is possible by flash photolysis[2] to provide a sufficiently high concentration of $^*UO_2^{+2}$ to show that it is dark blue with a broad absorption band close to 17000 cm^{-1} with some weak structure, presumably due to co-excited vibrations. It is striking that the five spectra reported[133–135] are slightly different, suggesting a weak interaction with the surrounding molecules in the ambient medium. We do not know the ε values, but it seems to be a very strong band (and hence parity-allowed) rather like the strong bands of brown I_3^- [39]. It is likely to represent the $\pi_g \rightarrow \pi_u$ transition, since $\sigma_g \rightarrow \pi_u$ at 17000 cm^{-1} would be expected to provide strong absorption bands $\sigma_g \rightarrow 5f$ close to 38000 cm^{-1} of the groundstate uranyl ion, if one supposes roughly additive M.O. energies, neglecting the effects of Eq. (2).

One of the more enigmatic problems is whether $^*UO_2^{+2}$ is bent or linear in its vibronic groundstate. The classical answer[22] is that the Stokes threshold coincides in the absorption and the emission spectra, indicating linear $^*UO_2^{+2}$. There is little doubt that the bending, if at all oçcurring, is not pronounced, though the co-excitation of the odd vibrational frequencies both in absorption and in fluorescence is slightly unexplained[22, 73, 94]. Whereas strong manifestations of the Jahn-Teller effect[32, 35] occur, when two M.O. having exactly the same energy in a given high symmetry are populated in a unbalanced fashion (by 1 and 2 electrons; or by 0 and 1 electron) encouraging spontaneous deviations from the high symmetry, the corresponding tendency in linear molecules containing 3 or more nuclei[62] is called the Renner effect. It is realized that an excited state of the uranyl ion containing two π_y electrons concentrated around the Cartesian y-axis, but only one π_x electron, may induce a bending, which may rotate symphonically with the formation of two orthogonal linear combinations of π_x and π_y. Such a rotation of the plane containing the three nuclei is the Renner analogy to the dynamic Jahn-Teller effect[136]. As a general comment, it may be said[32] that the absence of strong bending of $^*UO_2^{+2}$ shows that the chemical bonding is mainly taken care of by central atom orbitals having one kind of parity (e.g. U 6d) rather than the other (e.g. U 5f).

In the rest of this review, it has tacitly been assumed that fluorescence comes from the lowest excited state of the uranyl ion, a definite Ω level of Eq. (6), such as Ω = 1 of $UO_2Cl_4^{-2}$ [73] and that all of the higher Ω levels, of the same or of other M.O. configurations, decay by non-radiative processes to this level (or to the groundstate). Actually[22] there is very little evidence available for luminescence of higher levels. Nevertheless, as pointed out by Flint[137] one would normally observe weak luminescence from higher levels, if sufficiently sensitive photon counters are used. In condensed matter, Eq. (9) hardly allows radiative life-times below 10^{-8} s for transitions in the visible, and typically weak transitions are rather in the millisecond range if $P = 10^{-6}$. It is rare to determine non-radiative decay times more rapid than 10^{-11} s, but since the luminescence is a competing process starting immediately, quantum yields 10^{-3} or 10^{-8} would occur in the two cases. It is possible that non-observed luminescence with even lower quantum yield corresponds to non-radiative decay before 10^{-12} s but this is expected to be close to the conceivable limit. Though one may have endless difficulties with fluorescent trace impurities, the detection of such transitions is made much easier by time-resolved spectroscopy, if the emission can be studied within 10^{-7} to 10^{-6} s rather than under steady-state conditions. Anyhow, the only instance of higher-lying fluorescent levels of UO_2^{+2} reported by Hall and Dieke in $Rb_2UO_2Cl_4$ seems rather related to the 6 origins of $Cs_2UO_2Br_4$[138)] situated at 19622, 19628, 19638, 19653, 19662 and 19671 cm^{-1} with τ between 0.4 and 1.3 ms to be compared with 19665 and 19674 cm^{-1} reported by Flint and Tanner[246)] who note that this crystal has a strong tendency to develop luminescent traps at crystalline defects. This behaviour can be quite intricate at low temperature in crystals which allow extensive energy migration.

An old-known characteristic of $^*UO_2^{+2}$ is the *Becquerel effect,* that platinum (or other) electrodes immersed in illuminated and dark samples of the same composition show a potential difference. Since it is usually within the range 0.01 to 0.1 V, it has mainly academic interest, like several other photo-electrochemical potentials[89)]. There has been some discussion as to whether traces of organic molecules mediate this, not particularly reproducible, effect[22)].

Fayt et al.[248)] report a new band of UO_2^{+2} close to 560 nm growing up at high oxygen pressure (say 110 atm.). It is absolutely excluded that the excited state is an unknown triplet, but it may correspond to electron transfer between distinct molecules[249)] such as the purple colour of condensed mixtures[250)] of the colourless xenon and yellow gas IrF_6 (forming $Xe^+IrF_6^-$ in the excited state) and hence conceivably due to an excited state $O_2^+UO_2^+$.

7 Comparison with Uranium Hexafluoride

It is of great interest to compare the regular octahedral molecule UF_6 (which can also be studied in the gaseous state, the solid having a considerable vapour pressure) with the obstinately linear uranyl ion. The gaseous $5f^1$ system NpF_6 has[139)] two strong electron transfer bands at 38800 cm^{-1} ($\varepsilon = 1675$) and 45250 cm^{-1} ($\varepsilon = 2360$) whereas the $5f^2$ system PuF_6 has[140)] four broad bands at 31700 ($\varepsilon = 1055$), 40000 (1925), 44450 (2730) and 47850 cm^{-1} ($\varepsilon = 3685$) among which the first is the most clearly isolated. These bands are quite similar to the strong electron transfer bands[55)] of $5f^3$ neptunium(IV) and $5f^4$ plutonium(IV) hexahalides dissolved in acetonitrile (they tend to loose X^- in solvents[141)] with readily mobile protons). It is a general trend for electron transfer spectra[30,31,36)] to shift toward lower wave-numbers as a function of increasing atomic number Z, keeping the oxidation state and the ligands invariant. This trend is, in final analysis, due to the imperfect screening of the additional positive charge on the nucleus by the contracting electronic density of inner shells, making the central atom progressively more oxidizing (as long the loosely bound shell, receiving the electron from the ligands in the excited state, is not completely filled). The same phenomenon is observed in the chemical behaviour, e.g. the molecules AmF_6 and CmF_6 cannot be prepared.

Hence, one would extrapolate that UF_6 has a strong electron transfer band somewhere between 48 000 and 45 000 cm^{-1}, and indeed, two recent careful studies[142, 143] agree that a strong band occurs at 46 800 cm^{-1} with ε about 3000. However, they also agree that at least three weaker shoulders occur at 38 400 ($\varepsilon = 350$), 33 200 and 31 000 cm^{-1} and a first band split in two components at the surprisingly low wavenumbers 25 800 and 27 000 cm^{-1} ($\varepsilon = 6$ and 8). Further on[142, 143] this weak band fluoresces (both in the gaseous state and in cooled argon films) in the violet showing the Stokes threshold at 24 166 and 24 564 cm^{-1}, respectively. Like in the uranyl ion, the emission spectrum consists of a roughly equidistant vibrational progression[144] with characteristic frequencies, but mainly below 660 cm^{-1}. Another origin in cooled matrix-isolated UF_6 occurs at 30 331 cm^{-1} [143], and Grzybowski and Andrews[144] report two origins (at 12 K) at 24 641 and 24 649 cm^{-1}, possibly due to two distinct sites[144]. Repeated studies[251] show three sites in solid argon, but only one (at 24 612 cm^{-1}) in solid krypton. Crystalline UF_6 shows the origin[252] at 24 573 cm^{-1} at 1.6 K, and the mixed crystal $(UF_6)_{0.05}(WF_6)_{0.95}$ at 24 590 cm^{-1}.

These observations are very difficult to conciliate with the electron transfer spectra of other f-group hexahalide complexes. The first electron transfer band[43] of $4f^0$ cerium(IV), $4f^6$ europium(III) and $4f^{13}$ ytterbium(III) hexachloro and hexabromo complexes is ascribed to the parity-allowed transition from the loosest bound M.O. having the symmetry type $(\pi)t_{1g}$ in the point-group O_h to the 4f shell (of odd parity). The corresponding transition[35–39] in 4d and 5d group hexahalide complexes is far weaker than the parity-allowed transitions, but still has the lowest energy. It is tempting[1] to ascribe the weaker transitions of UF_6 and in particular the weakest band at 25 800 cm^{-1}, which seems at first incompatible with Eq. (1) and the mere existence of UCl_6[56], to traces of impurity molecules such as F_5UCl, F_5UOUF_5, F_5UOF, F_5UCF_3 or F_5USF_5 (of which analogs exist in sulphur(VI) chemistry) in view of the notorious chemical reactivity of UF_6. In this connection, we return to OUF_5^- [145] below, but the consistent studies[142, 143] of UF_6 (in particular the reproducible $\varepsilon = 6$ and 8 of the two first bands) would encourage the entirely revolutionary hypothesis that all uranium(VI) compounds have a specific band in the blue, if it was not for the enormous amount of facts[35, 36] available about electron transfer bands, though it must be admitted that d-group hexafluorides[146] may be slightly less well incorporated in a coherent pattern than corresponding complexes of heavier halides. However, it must also be remembered that impurities of aqua or hydroxo complexes do not provide electron transfer bands at smaller wave-numbers in the case of chlorides, bromides and iodides, as would be true in fluorides.

Four recent attempts to calculate one-electron energies in UF_6 by $X\alpha$ relativistic techniques[70, 147–149] provide a conspicuous dispersion of results in Table 3, much like the comparable quantities calculated for UO_2^{+2} in Table 1. At least, the neutral molecule UF_6 has no problem needing a background potential.

The two first columns give the non-relativistic and relativistic $X\alpha$ results (without the resolved spin-orbit coupling shown in the three last columns) by Boring and Wood[70] with the major origin (uranium 5f or fluorine π and σ orbitals) of the one-electron functions. These results are calculated in a self-consistent field as one-component (rather than four-

Table 3. M.O. energies (all in eV) calculated for the uranium hexafluoride molecule. When effects of spin-orbit coupling are neglected, the symmetry types t_{1u}, ... of the point-group O_h are indicated. The quantum numbers γ_6, γ_7 and γ_8 refers to the corresponding double-group (used for describing relativistic effects) in which case the parity can be seen from the main component of the one-electron function given at first in parenthesis

	non-relativistic $X\alpha$ (Ref. 70)	relativistic $X\alpha$ (Ref. 70)	Dirac-Slater (Ref. 147)		Dirac-Slater (Ref. 148)	ab initio (Ref. 149)
$(U\,5f)t_{1u}$	5.52	4.76	$(t_{1u})\gamma_6$	4.39	4.20	–
$(U\,5f)t_{2u}$	7.51	6.33	$(t_{1u})\gamma_8$	4.46	4.38	–
$(U\,5f)a_{2u}$	8.08	6.93	$(t_{2u})\gamma_7$	5.74	5.76	–
$(\pi + \sigma)t_{1u}$	9.73	10.58	$(t_{2u})\gamma_8$	5.97	5.89	–
$(\pi)t_{1g}$	11.28	11.24	$(a_{2u})\gamma_7$	6.75	6.89	–
$(\sigma)a_{1g}$	10.37	12.00	$(t_{1u})\gamma_8$	9.58	9.19	17.88
$(\pi)t_{2u}$	12.09	12.03	$(t_{1u})\gamma_6$	10.72	10.27	19.11
$(\sigma + \pi)t_{1u}$	12.58	12.60	$(t_{1g})\gamma_8$	10.68	10.25	18.55
$(\pi)t_{2g}$	12.57	12.61	$(t_{1g})\gamma_6$	10.72	10.28	18.57
$(\sigma)e_g$	13.74	13.93	$(t_{2u})\gamma_7$	11.35	11.10	19.63
			$(t_{2u})\gamma_8$	11.36	11.10	19.65
			$(a_{1g})\gamma_6$	11.67	11.06	19.73
			$(t_{1u})\gamma_8$	11.86	11.60	20.09
			$(t_{1u})\gamma_6$	11.97	11.59	20.16
			$(t_{2g})\gamma_7$	11.96	11.73	20.55
			$(t_{2g})\gamma_8$	12.03	11.82	20.59
			$(e_g)\gamma_8$	13.20	12.47	20.74
			$(t_{1u})\gamma_8$	22.45	–	35.20
			$(t_{1u})\gamma_6$	27.97	–	41.95
			$(a_{1g})\gamma_6$	29.51	–	44.50
			$(e_g)\gamma_8$	29.99	–	44.71
			$(t_{1u})\gamma_8$	31.06	–	46.04
			$(t_{1u})\gamma_6$	34.51	–	48.56
			$(a_{1g})\gamma_6$	51.13	–	70.92

component) relativistic functions. Case and Yang[147)] perform Dirac-Slater $X\alpha$ with effective exchange on four-component one-electron functions. Koelling, Ellis and Bartlett[148)] used the discrete variational method of Dirac and Slater, also on NpF_6 and PuF_6 obtaining slightly more negative one-electron energies than for UF_6 but without any dramatic differences. Hay et al.[149)] performed an ab initio Hartree-Fock treatment with a relativistic effective core potential. In this case, the one-electron energies do not have exactly the same significance as in $X\alpha$ techniques, and it is not surprising that their absolute values are some 8 eV higher than in the first four columns in Table 3. Actually, Hay et al.[149)] point out the excellent agreement between their values (rendered 3.9 eV less negative) and the photo-electron signals of UF_6 obtained by Karlsson et al.[150)] using 21.2 and 40.8 eV photons. The valence region of 12 different relativistic one-electron energies (calculated between 17.88 and 20.74 eV) corresponds to signals with I between 13.9 and (a rather broad vibrational structure centered at) 17.3 eV. By the same token, good agreement is obtained if the I-values of refs.[147)] and [148)] are increased by some 4 eV.

In many ways, the order of one-electron energies in Table 3 is much more similar in different calculations than for UO_2^{+2} in Table 1. General agreement exists that the lowest I among the filled orbitals belongs to the γ_8 component (two degenerate orbitals in the point-group O_h whereas γ_6 and γ_7 correspond each to one orbital) of t_{1u}. The spin-orbit splitting of t_{1u} is so strong that Hay et al.[149] situate the three t_{1g} orbitals in the interval between γ_8 and γ_6 of t_{1u}. However, all five columns of Table 3 agree that the barycentre of t_{1u} is situated above t_{1g}. This is a most striking difference from the excitation energies in all known 4d- and 5d-group hexahalide spectra[31, 35–39] which are always some 3000 cm^{-1} (0.4 eV) lower for t_{1g} than for t_{1u} electrons, and from the order of I-values obtained from photo-electron spectra of SF_6, SeF_6, TeF_6[151], MoF_6 and WF_6[152] where I is always the lowest for the t_{1g} linear combination of fluorine 2pπ orbitals which is non-bonding with respect to the central atom, but which may be strongly inter-ligand anti-bonding[32] because of the four node-planes. It may be noted that I = 13.9 eV (strictly speaking, the structure has the centre at 14.1 eV) for the γ_8 component of t_{1u} in UF_6[150] and even I = 15.3 eV (there is a shoulder at 15.1 eV) for t_{1g} are rather low, compared with t_{1g} having I = 15.7 eV in SF_6, 15.07 in MoF_6 and 15.35 eV in WF_6. There seems to be a general tendency for the lowest I in gaseous fluorides[48, 50] to decrease as a function of increasing size of the central atom.

The best established relativistic effects in chemistry are the high stability of 6s-like orbitals in thallium(I), lead(II) and bismuth(III) compounds[32, 41] corresponding to the finite asymptotic density of s-orbitals at the nuclear surface. This can also be observed in monatomic entities (such as the marginally higher ionization energies of Ra and Ra^+ compared with Ba and Ba^+). Whereas the finite amplitude of s-orbitals occurs already in the hydrogen atoms, the odd (j = 1/2) constituting a-third of the p-electronic density, is attracted by the nucleus (compared with j = 3/2) and has a density there approximately $(Z/137)^2$ times as large as the s-orbital under equal circumstances. Among the rare clear-cut $6p_{1/2}$ stabilizations in chemistry may be the properties of bismuth(I) in molten salts[32] but the $I(6p_{1/2})$ of the radon atom is 3.83 eV higher than $I(6p_{3/2})$ = 10.75 eV showing the beginning of the decline of nl-shell structure after Z = 80. Several reviews[153–155] of relativistic effects in chemistry concentrate on predicting chemical properties of elements with Z above 102[32, 156, 157] where the influence of deviations from the non-relativistic situation are significant, though they are much more spectacular on the energies of the inner shells. In the M.O. energies, the relativistic effects are also more pronounced than in the straightforward chemical behaviour.

Because the first-order spin-orbit coupling (following Landé's interval rule) can be readily incorporated in Russell-Saunders (S, L)-terms, it is frequently not considered as a relativistic effect (though it would disappear if the reciprocal value of the velocity of light in vacuo vanished). In UF_6, there is general agreement (Table 3) that γ_8 and γ_6 of the highest filled t_{1u} are separated 1.08 to 1.23 eV, i.e. 0.11 to 0.12 times the calculated separation[68] 9.6 eV between U $6p_{3/2}$ and U $6p_{1/2}$ in U^{+2} (mentioned in Table 1). Hay et al.[149] indicate that t_{1u} has 7% U p and 5% U 5f character (corresponding to the first of the empty 5f orbitals given in Table 3 being σ-anti-bonding). It is not said how much U 6p and U 7p participate in the p-character, but it is stated that U 7p is relatively much less important. In spite of 11% U 6d character of the filled t_{2g} orbital, γ_7 and γ_8 are only separated to the extent of 0.05 eV (Table 3) and in general, strong spin-orbit coupling is only computed for the four sets of filled t_{1u} orbitals.

The contraction of the inner s- and $p_{1/2}$ shells provides various indirect relativistic effects. Thus, 5f in the gaseous uranium atom[149] is destabilized by 6 eV compared to a non-relativistic wave-function. It is interesting to note that the average radius of the 5f shell[149] is close to 1.5 bohr (0.8 Å) and slightly smaller than 1.8 bohr of the 6p shell.

Closely similar results are obtained by Pyper and Grant[158] for multi-configuration Dirac-Fock treatment of gaseous ions U^{+} to U^{+5}. The condition for the average energy of the electron configuration $a^v b^w$ to be situated at the mean value of $a^{v+1}b^{w-1}$ and $a^{v-1}b^{w+1}$ is[31, 159] that J(a, b) in Eq. (2) is the arithmetic average of J(a, a) and J(b, b). Recently, Wyart, Kaufman and Sugar[160] analyzed the *J*-levels of gaseous U^{+4} and found the configuration [80] $6p^5 5f^3$ to start 145871 cm^{-1} above the groundstate belonging to [86] $5f^2$. Their Hartree-Fock result that the configuration barycenters $6p^6 5f^2$, $6p^5 5f^3$ and $6p^4 5f^4$ are approximately equidistant corresponds to the roughly equal average radii of the 5f and 6p shells.

For our purpose, the main conclusion of Table 3 is that the highest filled t_{1u} is "pushed from below" by U 6p like the effect of N 2s on the loosest bound M.O. of N_2 (having the symmetry type σ_g) discussed in Sect. 3. It does not seem likely that the admixture of fluorine π and σ character in t_{1u} which has quite spectacular effects on the relative intensities of electron transfer bands in d-group hexahalide complexes[36, 37] would be sufficient to explain the low I (even in a non-relativistic treatment). Seen from this point of view, 4f group hexahalides[43] of Ce(IV), Eu(III) and Yb(III) shows no weak absorption bands due to $t_{1u} \rightarrow 4f$ at lower wave-numbers than the first strong transition $t_{1g} \rightarrow 4f$ whereas UO_2^{+2} may represent an analogous case of "pushing up" of σ_u by 6p (though Table 1 does not give conclusive evidence for the order of excitation energies proposed by Denning) though it is slightly worrying that Boring and Wood[70] indicate 71% U5f character of the highest filled σ_u in the non-relativistic Xα treatment of the uranyl ion, and 59% U5f in the relativistic orbital calculated at 23.10 eV in the last column of Table 1. This predominance of U5f looks more like the paradoxical effects[46, 161, 162] of certain anti-bonding orbitals in a partly filled shell having higher I (though much lower electron affinity) than their bonding counter-parts of the same symmetry type.

Recently, Pyykkö and Lohr[255] applied a relativistic extended Hückel model to several heavy species including UO_2^{+2}, $UO_2Cl_4^{-2}$, UF_6 and UCl_6. The uranyl I-values 10 eV for σ_u, 12 eV (with appreciable spin-orbit coupling) for π_u and about 13 eV for the M.O. of even parity agree even better with Denning's hypothesis than the results of Walch and Ellis[69] in Table 1. However, the squared 5f amplitude of σ_u is 86%, making a violent contrast to "ligand field" effects in lanthanides[32, 220] by the first excited state of the uranyl ion being essentially the transition from a bonding to a non-bonding 5f-like orbital.

Even when it is accepted[147–149] that the γ_8 component of t_{1u} in UF_6 has I 0.7 to 1.1 eV lower than t_{1g} it still does not solve all the problems. In order to understand the electron transfer spectrum of this molecule, it is useful to consider the $5f^1$ systems UF_6^- [67] and NpF_6[139] having quite sharp transitions (in eV):

	Γ_{7u}	Γ_{8u}	Γ_{7u}	Γ_{8u}	Γ_{6u}	
UF_6^-	0.00	0.57	0.86	1.58	1.77	(11)
NpF_6	0.00	0.94	1.18	(2.82)	(2.97)	

The order of γ-values for the empty M.O. of UF_6 is in perfect agreement with Table 3, and the over-all spreading of energies 2.36 eV[147] or 2.69 eV[148] is quite compatible with Eq. (11). The five Γ_J values play the same rôle as the seven Ω-components of $^2F_{5/2}$ and $^2F_{7/2}$ used for description of the electron transfer states of a linear uranyl ion. However, the number of Γ_J values of the excited states is quite large[148] because transfer of t_{1u} or t_{1g} in each case provides 84 states.

In our opinion, the most serious problem in UF_6 is that the two weak bands at 25 800 and 27 000 cm^{-1} (if ascribed to excitation of the γ_8 component of t_{1u} according to Table 3) is followed by a lot of rather indeterminate, and almost equally weak bands (which may readily represent transitions to the many states implied by Eq. (11)) but that the first band having ε above 50 is the shoulder at 38 400 cm^{-1} ($\varepsilon = 350$). Experience from 4f group hexahalides[43], NpX_6^{-2} and PuX_6^{-2}[55] suggests that the allowed $t_{1g} \rightarrow$ 5f transitions should have ε at least above 2000, and from this point of view, the first band of UF_6 that fits the bill occurs at 46 800 cm^{-1} with ε about 3000. In other words, we have a 2.5 eV wide wilderness of exceptionally weak electron transfer bands, whereas the gap between γ_{8u} and t_{1g} expected from Table 3 is 1 eV. It is true, as pointed out by McDiarmid[142] that the transition from t_{1g} to the lowest 5f orbital γ_{7u} does not have an electric dipole moment to the extent that the symmetry type of γ_{7u} is a_{2u} but it is very difficult to keep such forbidden transitions down at very low ε values. To summarize: we are stuck with the problem that the Stokes threshold of UF_6 is situated at 3.00 eV, the first strong band maximum occurs at 5.8 eV, to be compared with a weak precursor band ($t_{1g} \rightarrow$ (d)t_{2g}) in MoF_6 at 5.9 eV and in WF_6 at 7.2 eV[152] and a strong band at 6.54 and 8.05 eV, respectively. It is argued by McDiarmid[142] that the corresponding strong transition to U6d occurs (with $\varepsilon = 20 000$) in UF_6 at 156 nm (64 100 cm^{-1} = 8.0 eV). There is nothing impossible in excitation to U6d needing 5 eV more energy than to U5f but it would stabilize UF_6^- tremendously if U5f is that more bound. It might be argued that the origin at 3.00 eV corresponds to an entirely differing disposition of the seven nuclei than in the groundstate of UF_6. However, the Stokes shift of luminescence[144, 251, 252] is not extremely large, though the t_{2g} vibrations are excited to an extent suggesting strong deviations of the excited state from equilibrium cubic symmetry.

Koelling, Ellis and Bartlett[148] expect a large number of weak electron transfer bands of NpF_6 and PuF_6 at slightly lower wave-numbers than found in UF_6. At the beginning of this section, we discussed the strong transitions, but it is beyond doubt that if UF_6 has genuine transitions at 25 800 and 27 000 cm^{-1}, one would expect comparable features in NpF_6. Steindler and Gerding[139] report very weak features at 22 700, 23 950, 24 600 and a doubtful shoulder at 26 000 cm^{-1} with the ε values 1.5, 2.7, 4.2 and 9, respectively. The reason why we put parenthesis in Eq. (11) around the energies 22 700 and 23 950 cm^{-1} is that they may represent electron transfer bands (with a world lower record of $\varepsilon = 2$) rather than transitions to the σ-anti-bonding γ_{8u} and γ_{6u} orbitals. The situation is chaotic in PuF_6 with very crowded $5f^2$ transitions of comparable (and higher) intensity. We return in Sect. 10 to the related problem in PuO_2^{+2} and AmO_2^{+2}. Thornton, Rösch and Edelstein[163] performed quasi-relativistic self-consistent field Xα studies of nine $5f^1$ hexahalide complexes and found $(\pi + \sigma)t_{1u}$ slightly above $(\pi)t_{1g}$ in PaF_6^{-2} and UF_6^- whereas NpF_6 and PaX_6^{-2} and UX_6^- (X = Cl, Br, I) should show the opposite order. Reasonable agreement was obtained with the energy differences between the 5f-like orbitals a_{2u}, t_{2u} and t_{1u}.

Thornton et al.[164] have taken up the problem of gaseous UCl_6. They determine the photo-electron spectrum with 21.2 and 40.8 eV photons, finding I = 11.28, 11.74, 12.07, (12.34), 12.45 and 13.28 eV of six signals (with a shoulder in parenthesis). In agreement with photo-electron spectra of SF_6 and similar molecules, and with electron transfer spectra of d-group hexahalides, the lowest I-value is ascribed to $(\pi)t_{1g}$ and the next-lowest to $(\pi + \sigma)t_{1u}$. The shift from the lowest I = 14.1 eV in UF_6[150] is unusually modest for a chloride (normally, the shift is about 4 eV). These authors also performed a discrete variational Xα calculation, predicting I-values for 24 valence electrons between 11.60 eV (t_{1g}) and 13.46 eV (e_g). The spin-orbit coupling was not introduced. Reliable data for the visible absorption spectrum of the dark-coloured UCl_6 in any physical state are still painfully lacking.

For completeness, we may mention that the detailed positions of absorption bands of UF_6 have been essentially confirmed via electron energy loss spectra between 3 and 28 eV[165] with very strong peaks at 12.4 and 13.1 eV and quite strong peaks at 5.8, 7.9 and 9.2 eV. A comparative study of the electron-impact excitation spectra[166] of UF_6 and WF_6 have confirmed the weaker peaks of UF_6 at 3.3 and 4.2 eV and the stronger at 5.8, 6.9, 8.0, 9.3 and 11.0 eV, followed by sharp, intense peaks at 12.6 and 13.2 eV. One of the advantages of this technique is that the peak intensities do not differ by several orders of magnitude, as do the absorption bands. It was confirmed[166] that WF_6 has no energy loss below 5.9 eV, three strong peaks at 7.1, 7.9 and 8.5 eV and two other strong peaks at 12.6 and 13.6 eV.

Quite recently, Cuellar and Marks[167] studied the red methoxide $U(OCH_3)_6$. Unfortunately, the solution in CH_2Cl_2 does not have a very clear-cut spectrum. The lowest-energy feature is a shoulder at 590 nm with $\varepsilon \sim 200$, another at 430 nm ($\varepsilon \sim 1000$) and a third shoulder at 350 nm ($\varepsilon \sim 10000$). The most constructive result is that this spectrum is the higher limit, since nearly all impurities have approximately additive spectra, but not negative contributions (unless solvent molecules perform rearrangement or coordinative reactions). The spectrum seems quite different from UF_6, the strongest band being shifted 18000 cm^{-1} toward lower energy, in the methoxide, suggesting $x_{opt} = 3.3$ for the latter ligand. The earliest and weakest band only shifts 9000 cm^{-1}. By performing proton and fluorine nuclear magnetic resonance[167] not only the five mixed complexes $U(OCH_3)_nF_{6-n}$ were detected, but also the two geometrical isomers for each of the cases n = 2, 3 and 4.

8 Comparison with Octahedral Uranate Chromophores

Though only five halides, UX_6 (X = F, Cl) and OUX_5^- (X = F, Cl, Br), are known to contain less than two oxygen atoms per uranium(VI), it should not be concluded that all solid mixed oxides of this oxidation state contain uranyl ions. Many of the solids precipitated from solution are amorphous, and others have not had their crystal structures determined. Many of the stoichiometric formulae given in literature are idealized simplifications (like for minerals) and there is little doubt that the commercial chemical (water-insoluble) $Na_2U_2O_7$ is very different from the dichromates, and much more similar to the dekavanadate $Na_6V_{10}O_{28}$. An interesting compound is the tetrahydrated peroxide $UO_2(O_2)$ precipitated from neutral solutions of uranyl salts with H_2O_2. Its lemon-yellow colour forms a striking contrast to the tomato-red peroxo complexes in acidic solution of Ti(IV), V(V) and yellow peroxo complexes of Mo(VI), when it is remembered[36] that x_{opt} of the peroxo ligand must be quite close to 2.8 for bromide. Anyhow, the mixed oxides of U(VI) with colourless cations are rather uniformly yellow to orange, and since they do not frequently fluoresce, they have not attracted much spectroscopic interest.

However, the *cubic perovskites* contain (at least on a time-average picture) a regular octahedron of six oxygen atoms surrounding U(VI). The chronological prototype of this crystal structure is $SrTiO_3$ with the tetrakaidecahedral (cuboctahedral) coordination number $N = 12$ for Sr(II) like in cubic close-packed metals, and regular octahedral $N = 6$ for Ti(IV). It is noted that this solid is a mixed oxide, and not strictly a titanate of strontium, since it does not contain discrete anions, as is the case for $SrCO_3$, YPO_4, $ErVO_4$ and $ZrSiO_4$(6). The tungsten bronzes such as Na_xWO_3 and La_xWO_3 are disordered perovskites in the sense that they have all titanium sites of $SrTiO_3$ occupied by tungsten atoms, but only a few of the strontium sites occupied by sodium or lanthanum. The metallic compound ReO_3 has a related cubic structure with all the Sr sites empty. Mixed perovskites containing two or three differing elements on the octahedral sites have played an important rôle in the study of "ligand field" transitions in d-shells[31, 136)]. In particular, it is noted that all the atomic positions in cubic perovskites are *special,* and hence, all internuclear distances are geometrical coefficients times the unit cell parameter a_0. The same is true for the cubic binary structures CsCl ($N = 8$), NaCl ($N = 6$) and CuCl ($N = 4$) (frequently called "zinc blende" but ZnS has also another, hexagonal, modification wurtzite, being the normal structure of BeO and ZnO with local $N = 4$). Not all atoms in all cubic structures are on special positions; the chlorine atoms in K_2PtCl_6 are on general positions, and the Pt-Cl distance represents one degree of freedom independent of a_0. Anyhow, decreasing a_0 (or the several unit cell parameters in non-cubic structures) is a convenient technique for imitating very high hydrostatic pressures (in the 10^4 to 10^5 atm class) as well-known[31)] for the red ruby $Al_{2-x}Cr_xO_3$ compared with the green isotypic Cr_2O_3. Expanding the crystal lattice (substituting with larger cations) can be even more helpful, because one cannot in practice realize negative hydrostatic pressures below -100 atm without cavitation breaking up the solid. Though one determines the unit cell parameter of a mixed-oxide perovskite, it is not perfectly excluded that the larger M pushes the oxygen atoms, and the smaller M′ pulls the oxygen atoms closer in the sequence MOM′OMOM′O ... of nuclei situated on a straight line. However, much spectroscopic evidence suggests that the oxide mobility is rather attenuated when M or M′ represents a small minority.

Blasse[58, 60)] and Bleijenberg[168)] reviewed the studies of uranium(VI) substituted in cubic perovskites of the kind $BaMg_{0.5}W_{0.5}O_3$. Whereas tungstates are tetrahedral, monomeric anions (with the first absorption band in weakly alkaline solution at 50300 cm^{-1}) known from crystals such as scheelite $CaWO_4$ or $PbWO_4$, we have here octahedral coordination, and there is an electron transfer band in the near ultra-violet at 32000 cm^{-1}. As frequently true for the formation of mixed oxides, the ionic size (here Mg(II) and W(VI) are comparable) is more important than the oxidation state. Electroneutrality of the crystal is maintained by appropriate stoichiometric proportions of differing oxidation states, or otherwise, as frequent in fluorite-type mixed oxides [6, 9)]by anion excess ($Ca_{0.8}Y_{0.2}F_{2.2}$) or, more frequently, by anion deficit (such as $Er_{0.5}Zr_{0.5}O_{1.75}$ or $La_{0.54}Th_{0.46}O_{1.73}$ still isotypic with ThO_2) In all such cases, we write the *sum* of all the atoms fully occupying a definite site as an integer, whereas we would write the cubic pyrochlore type $Er_2Ti_2O_7$ with conventional coefficients.

It is not surprising that the highly different oxidation states Mg(II) and W(VI) in spite of their comparable ionic radii form a systematically ordered, cubic superstructure of perovskite Ba_2MgWO_6. Hair and Blasse[169] introduced traces of U(VI) in such perovskites $E_2E'W_{0.997}U_{0.003}O_3$ and found at liquid nitrogen temperature quite long life-times (in μs) with E_2E' =

$$Ba_2Mg:255\ \ Ba_2Ca:240\ \ Ba_2Zn:260\ \ Ba_2Sr:200\ \ Sr_2Mg:245\ \ Ca_2Mg:220. \qquad (12)$$

Only in $SrLaNaW_{0.997}U_{0.003}O_3$ was τ found below 155 μs. The luminescence consists of a complicated vibrational structure with one or two extremely sharp lines of origin found (in cm^{-1}) by Bleijenberg and De Wit[170] at liquid helium temperature in $E_2E'W_{0.999}U_{0.001}O_3$ with E_2E' =

Ba_2Zn:	19090 & 19130	Ba_2Mg:	19980	
Ba_2Cd:	19280	Ba_2Ca:	20200	(13)
Ba_2Ba:	19670 & 19770	Ca_2Mg:	20220	
Ba_2Sr:	19900 & 19970	Sr_2Mg:	20330	

As in many other instances of luminescence it is difficult to tell whether the two origins correspond to two sites having comparable concentration, or whether the lower of the two wave-numbers is emitted by a very low concentration of energy traps at a rare defect site. It is beyond doubt that the U(VI) luminescence is pumped with energy transfer[33, 60] from the abundant W(VI). The ordered perovskites are not containing any uranium sharing its six ligating oxygen atoms with another U atom since the Cartesian axes contain alternatingly E'OWOE'OWO ... and in this sense, we have an octahedral anion UO_6^{-6} like the monomeric perxenate XeO_6^{-4} though in the perovskite structure, each oxygen atom is bound to two of the small cations (one E' and one W or U) with $N = 6$ and, in a plane, to four cations (E) with $N = 12$, and hence, oxygen is situated in a tetragonally distorted octahedron. The ordered W(VI) perovskites such as Ba_2MgWO_6 have a strong electron transfer band close to 32000 cm^{-1}. This behaviour of WO_6^{-6} was analyzed with (non-relativistic) self-consistent M.O. calculations by Van Oosterhout[171] concluding in the conventional order of excitation energies $(\pi)t_{1g} < (\pi + \sigma)t_{1u} \sim (\pi)t_{2u}$ ... well established[36, 37] for WCl_6 and[172] the isoelectronic $TaCl_6^-$ and isologous $NbCl_6^-$. One might have expected that the Stokes thresholds in Eq. (13) within 1 to 3% having the same positions as for typical uranyl salts might indicate a distorted UO_6^{-6} chromophore on an instantaneous picture[32] with two short U-O distances on one Cartesian axis and four long distances on the two other axes. Such a distortion is the opposite of the copper(II) distortion related to the Jahn-Teller effect[7, 136] where a typical example is the NaCl-structure of $Cu_xMg_{1-x}O$ for $x < 0.08$ (black tenorite CuO with quadratic coordinated $N = 4$ starts precipitating at higher x values) where the unit cell remains constant for $x < 0.05$ shows the perfect cubic structure (time-average picture of the average unit cell) though the $3d^9$ absorption bands in the near infra-red still show strong deviations from having three equivalent Cartesian axes on an instantaneous picture[173]. There are indications from crystal structures of this tendency in $MgUO_4$ (isotypic with $MgWO_4$ and not with scheelite $CaWO_4$) where a tetragonally distorted octahedron with two U-O distances 1.92 Å and four long distances (2.16 to 2.20 Å) to oxygen atoms in the equatorial plane shared with another octahedron approaches the situation in a uranyl complex. The luminescence[174] produced by defective traps has $\tau = 0.1$ ms and shows typically six origins (due to differing sites) between 17920 and 18090 cm^{-1} at lower energy than the regular octahedra in Eq. (13).

Anyhow, the tetragonal distortion does not seem the problem in the ordered perovskites. The vibrational structure associated in luminescence (partly excited by highly monochromatic green light) with the origin at 19980 cm^{-1} in $Ba_2MgW_{0.999}U_{0.001}O_3$ can be analyzed[168, 170] in terms of the characteristic frequencies of the O_h symmetry of the groundstate UO_6^{-6}. Actually, the emission spectrum at 4 K is not so much a progression of the totally symmetric "breathing" frequency 797 cm^{-1} (as in many solids, the anharmonicity has the opposite sign to that usually found in diatomic molecules in the excited state, where the excitation spectrum of UO_6^{-6} shows ν_1 at 724 but 2 ν_1 at 1531 cm^{-1}) as dominated by the "bending" frequencies at 349, 460 and 287 cm^{-1} having the symmetry types t_{1u}, t_{2g} and t_{2u}, respectively. However, the moderate Stokes shift and decrease of ν_1 (relative to the electronic groundstate) does not suggest the potential surface of the luminescent state to have a distribution of the seven nuclei extremely different from the groundstate.

There is no doubt that the luminescence is not fully allowed as an electric dipole transition. The life-times in Eq. (12) of the order 0.3 ms are hardly compatible with ε above 20 of the corresponding absorption band, when Eq. (9) is carefully considered. Hence, it cannot be due to electron transfer to U 5f from an even orbital such as t_{1g}. Bleijenberg[168] assumes the situation to be analogous to UF_6 (isoelectronic with UO_6^{-6}) and the first excited state corresponding to transfer from the γ_8 component of t_{1u} (cf. Table 3) to the first empty 5 fγ_7 orbital. By the way, the 8 mutually orthogonal states having the symmetry types $\Gamma_J = E_g, T_{1g}, T_{2g}$ (like for even $J = 1$ and 2) in O_h would be expected to provide adjacent energy levels close to the luminescent level. Hence, the Stokes threshold of Eq. (13) occurs only 4200 cm^{-1} (or 0.14 units of optical electronegativity) below that in UF_6 showing once more that electron transfer from the odd γ_8 component does not follow the usual regularities of x_{opt}. Ref. 168, p. 105 compares the theoretical predictions[70, 148, 166] of optical transition energies in UF_6 with UO_6^{-6}.

It may be noted that $Li_6W_{0.997}U_{0.003}O_6$ at the same time[175] emits the broad-band emission characteristic for WO_6^{-6} and, at lower wave-numbers, the green lines of UO_6^{-6} with the origin at 19970 cm^{-1}.

In this situation, the absorption and excitation bands at higher wave-numbers in the U(VI) containing perovskites are highly interesting. At room temperature, $Ba_2MgW_{0.997}U_{0.003}O_3$ has an excitation spectrum showing a strong band at 30000 cm^{-1} just before the strong features intrinsic to W(VI), and a weaker band at 26500 cm^{-1} which becomes structured at low temperature. As discussed in further detail by Bleijenberg[168] this band is likely to be a parity-allowed transition from $(\pi)t_{1g}$ to U 5 f.

There is a pronounced thermal[176, 177] and concentration quenching[176] for x above 0.01 in $Ba_2MgW_{1-x}U_xO_6$. One origin of this quenching is energy migration[174, 176] to U(V) traces. The absorption spectrum of UO_6^{-7} in perovskites shows narrow absorption bands[178] in the near infra-red, among which the highest between 7320 and 7410 cm^{-1} is analogous to the peak at 9530 cm^{-1} in the isoelectronic NpF_6[139] or 7410 cm^{-1} in UF_6^-[179], and due to the transition between the two Γ_{7u} levels in Eq. (11). Like in UX_6^{-2}[55, 141] there is extensive co-excitation of vibrations in these transitions. The photo-electron spectra of U(V) in mixed oxides show quite characteristic features[180].

One of the most successful techniques of detecting small amounts (say 10^{-9} to 10^{-6} g) of uranium is the fluorescence after incorporation in molten sodium fluoride[22, 181]. Since the light emitted is green, and since the emission spectrum shows some vibrational structure, it was the general opinion before 1956 that some kind of imbedded uranyl fluoride was formed. However, it is clear from our discussion above that one cannot distinguish linear uranyl groups from octahedral uranate groups only by the position of the Stokes threshold. Thus, the highly anisotropic uranyl-like chromophores in $MgUO_4$[174)] have their electronic origins at 10% lower wave-numbers than typical uranyl salts, whereas the cubic perovskites in Eq. (13) have origins almost coinciding with the values in Table 2. The first to propose the existence of $U(VI)O_6$ chromophores in NaF was Runciman[182)] based on experimentation involving ^{18}O isotopic shifts of the vibrational structure. Further work clearly demonstrated four kinds of luminescent U(VI) in NaF, as reviewed by Bleijenberg[168)]. One centre formed preferentially in the presence of readily available oxygen is indeed UO_6^{-6} characterized by the strongest emission line (at 4 K) at the relatively low photon-energy corresponding to 563.6 nm. Since this anion replaces a building-block NaF_6^{-5} with one negative charge less, it should be accompanied by a fluoride-vacancy in the crystal. The spectrum having the strongest line at 547.2 nm belongs to UO_5F^{-5} which can replace NaF_6^{-5} without charge compensation. Under reducing circumstances, a group $UO_4F_2^{-4}$ needing the presence of a Na^+ vacancy can be studied, with the strongest line at 551.5 nm. Finally, there is a dimeric species said[168, 183)] to consist of two UO_6^{-6} bound together, with the strongest line at the shortest wavelength 540.8 nm. At this point, we would like to mention that $O_4UO_2UO_4^{-8}$ having two oxide bridges shared by two octahedra (like the two chloride bridges in crystalline Nb_2Cl_{10}) can replace $Na_2F_{10}^{-8}$ without any charge compensation. The detailed behaviour of these four centers have been elucidated with selective laser excitation[168, 183)] and the UO_6^{-6} (with adjacent F^- vacancy) shown to have two electronic origins, one at 552.8 nm (18089 cm^{-1}) and the other (with very long τ = 2.7 ms) at 563.6 nm (17743 cm^{-1}). The argument runs that the local symmetry is decreased from O_h to C_{4v} (or some other uniaxial point-group). Lys and Runciman[184)] studying absorption and emission spectra of U(VI) in several crystals such as rutile-type MgF_2 (where the Stokes threshold occurs at 18808 cm^{-1}) argue that the 563.6 nm line in NaF is no resonance line, since it could not be detected in absorption. After all, the 17743 cm^{-1} origin may be due to a long-lived trap with a specific choice of charge compensation.

Runciman and Wong[185)] performed selective laser excitation of traces of U(VI) dissolved in the cubic crystal LiF and found several centers, among which a prominent one has its Stokes threshold at 527.72 nm or 18944 cm^{-1}. By X-ray or γ irradiation[186)] it is possible to add an electron permanently to UO_5F^{-5} and to obtain the U(V) chromophore UO_5F^{-6} having characteristic electron paramagnetic resonance and optical spectra. At 77 K, the U(VI) in LiF[185)] has two strong absorption bands at 248 and 302 nm with oscillator strength P of the order of magnitude 1 and 0.1. At 77 K, τ for the luminescence is 1.4 ms and increases only moderately to 1.8 ms at 4 K.

In many ways, the luminescent group UO_5F^{-5} with symmetry C_{4v} is a pendant to the yellow UOF_5^- characterized by Bagnall et al.[145)]. For our purposes, it is interesting to compare with the red $UOCl_5^-$ with an electron transfer band at 24000 cm^{-1} indicating the optical electronegativity x_{uncorr} of Eq. (1) to be 2.2 for U(VI) in this case. Bagnall and collaborators[145)] also prepared the $P(C_6H_5)_4^+$ salt of the dark brown $UOClBr_4^-$ having an electron transfer band at 17200 cm^{-1} indicating a comparable x_{uncorr}. When comparing with the value 1.8 derived from the electron transfer band of $UO_2Br_4^{-2}$ at 29800 and of red $UO_2I_4^{-2}$ at 19200 cm^{-1} it can be concluded that the second oxo ligand renders U(VI) 0.4 unit or 12000 cm^{-1} less oxidizing (in the equatorial plane with long U–X distances). The effect of one oxo ligand replacing a halide was discussed by Ryan[179)] in the case of green $UOBr_5^{-2}$ having the first electron transfer band at 26700 cm^{-1} and the dark brown UBr_6^- at 17400 cm^{-1} (followed by a second band at 20000 cm^{-1}). By the same token[145], blue $UOCl_5^{-2}$ has the first electron transfer band at 32600 cm^{-1} to be compared with orange UCl_6^- [179] at 25300 cm^{-1}. In these four cases, x_{uncorr} for U(V) is 2.2 in the hexahalide complexes, and 1.9 when one oxo ligand is introduced. It is tempting to extrapolate that the optical electronegativity of U(VI) in a hexahalide would be 2.6, forming a linear variation with UOX_5^- and $UO_2X_4^{-2}$. In that case, the first strong electron transfer band would occur at 39000 cm^{-1} in UF_6 and at 12000 cm^{-1} in UCl_6. Even this assumption of strong "bleaching" influence of oxo ligands (as seems to occur[36] also for hydride and carbanion ligands) does not fit with $x_{uncorr} = 2.6$ for Np(VI) in NpF_6 and 2.85 for Pu(VI) in PuF_6 which would rather extrapolate to 2.4 for U(VI). In all cases, the ultra-weak electron transfer bands of UF_6 at 25800 and 27000 cm^{-1} remain mysterious.

The many instances of UO_6^{-6} in ordered perovskites and NaF discussed above may invite the comment that the Stokes threshold between 18000 and 20000 cm^{-1} might suggest x_{uncorr} of U(VI) to be about 0.7 unit lower than x_{opt} of the oxide ligand. Unfortunately, it is well established[59, 187] that oxide does not have an invariant x_{opt} but looks much more reducing in some compounds than in others. In our case, it can only be expected to fall somewhere in the interval between 3.5 (known for aqua ions) and 3.1 (valid for most oxides with a reasonably strong Madelung potential). This brings us to a rather wide range between 2.4 and 2.8 for U(VI) in UO_6^{-6}. Now, taking the other way, from the anomalous first band of UF_6, assuming U(VI) to remain equally oxidizing, oxide winds up with x_{opt} close to 3.7. The conclusion is perhaps that the whole concept of optical electronegativities works remarkably well within the realm of hexahalide complexes[30–44], reasonably well with a number of other ligands, but fails for oxo complexes. It is not even perfectly certain whether we have two distinct strong effects (oxo ligands are strongly π-anti-bonding on d-like orbitals; and oxo ligands make a central atom in a given high oxidation state look less oxidizing than it would have done under other circumstances) or whether they are two sides of the same coin. We must remember that the spectroscopic or other observable effects of covalent bonding in lanthanide compounds[6, 9] are extremely difficult to ascertain, if the L.C.A.O. model involves the empty 5d or 6s orbitals (where the quite small extent of the nephelauxetic effect provides a higher limit[31, 33] to the involvement of 4f orbitals). The same is true for the empty 6d and 7s orbitals in compounds of transthorium elements. The additional feature is the influence of relatively loosely bound closed shells, such as $6p_{3/2}$ playing a most unexpected rôle in the calculations on UF_6 summarized in Table 3. Said in other words, the presence of U 6p makes four of the 36 (or 48) valence electrons of the fluoride ligands particularly reducing.

The luminescence of U(VI) in many mixed oxides has been investigated. Thus, Blasse[188] studied the superstructures of NaCl represented by Li_3NbO_4, Li_3SbO_4, Li_4WO_5, Li_2SnO_3 and $LiScO_2$. The octahedral sites of incorporated Cr(III) or Rh(III) are known[188, 189] to provide sub-shell energy differences Δ (between the two strongly anti-bonding d-like orbitals and the three, roughly non-bonding, 3d or 4d orbitals) much smaller than in the corresponding $M(OH_2)_6^{+3}$ whereas $Al_{2-x}M_xO_3$ gives higher Δ values than the aqua ions. Traces of Cr(III) in such mixed oxides are known to quench U(VI) fluorescence very effectively. The luminescence of Cr(III) in glasses present enigmatic problems[190–192] of a certain technological importance[25].

9 Uranyl Glasses

Glasses used for incorporation of transition-group ions are mixed oxides (or in several recent studies, mixed fluorides or mixed sulphides) of a quite high number of elements. It is pragmatically known that the tendency toward crystallization (de-vitrification) of the glass is the least pronounced, when several elements are present, and when their relative concentrations are far removed from simple stoichiometric ratios. This rule has exceptions; dehydration of NaH_2PO_4 provides a phosphate glass with stoichiometric coefficients $NaPO_3$ like the many oligomeric metaphosphates, and silica SiO_2, yellow S_8 and red Se_8 can remain vitreous after having been molten. Weyl[193] wrote a fascinating book "Coloured Glasses" about the colours and absorption spectra (at a primitive level of instrumentation) produced by d- and f-group ions.

Between 1898 (when radium was discovered) and 1942 (when Fermi started the first uranium-graphite reactor) the main purpose of exploiting uranium minerals[194] was to extract the 0.3 g Ra in radioactive equilibrium with each ton U, and it was very difficult to sell uranium. Besides minor uses, such as the olive-green pigment U_3O_8 for porcelain painting, a relatively more important was the fabrication of conventional calcium lead silicate glasses made green fluorescent (and yellow colour in transparence) with uranium salts. Comparable emission spectra are observed[65] in $NaPO_3$ glass, where a Stokes threshold line is observed at 20530 cm^{-1}, accompanied by roughly equidistant components at 19710 (the most intense), 18840, 17950 and 17070 cm^{-1}. Almost the same emission spectrum is observed of a uranyl solution in 85 weight % phosphoric acid. By the way, the composition of this solvent is not clear-cut; besides $OP(OH)_3$, protonated forms of pyrophosphate $O_3POPO_3^{-4}$ and of various metaphosphates also establish equilibria slowly. The intensity of the fluorescence (not divided by the uranyl concentration) goes through a shallow maximum between 1 and 2 weight % UO_2^{+2} in the phosphate glass, corresponding to approximately 0.1 molar concentration. This concentration quenching does not involve collision complexes (as in water and other solvents of low viscosity) but may be related to cross-relaxation[33] like in the case of neodymium(III). The decay curve can be analyzed in the sum of two exponential curves[65], about a-quarter of the initial intensity decaying with $\tau = 115$ μs and the major part with $\tau = 367$ μs. The short-lived component may be due to pairs with short U–U distances formed during the cooling of the glass, and perhaps comparable to the exciplexes of Marcantonatos[101, 102]. However, even the short-lived component has the life-time about 100 times longer than of the uranyl ion in aqueous solution, though only about half the values for UO_6^{-6} (at 77 K) given in Eq. (12). At low uranyl concentration, the decay[65] becomes exponential with $\tau = 0.41$ ms.

It is not all glasses that show fluorescence spectra indicating the presence of triatomic uranyl groups. Both borate glasses (which can be made by melting borax $Na_2B_4O_7$) and germanate glasses (where silicon in conventional glasses is replaced by germanium) show a broad, weakly structured, emission band with maximum[33, 65] at 19000 and 18800 cm^{-1}, respectively, though in both cases[65] the Stokes threshold can be preceived as a shoulder at 20300 cm^{-1}, and $\tau = 0.15$ ms at room temperature. The form of the emission spectra is quite reminiscent of UO_6^{-6} in ordered perovskites, but an alternative explanation would be a superposition of two or three series of uranyl-like, approximately equidistant, progressions with a mutual shift of 300 cm^{-1}. This kind of problem is most readily studied with selective laser excitation, such as the various luminophores in NaF[183] and LiF[185]. Lam and Leung[195] studied soda-lime (sodium calcium silicate) glass excited by almost monochromatic light, and found strong variation of the emission spectrum as a function of the exciting wave-length, particularly between 510 and 525 nm. Among several electronic origins, the two most prominent were found at 516 nm (19374) and 506.5 nm (19738 cm^{-1}).

Recent work[196] on mixed borate-phosphate glasses show essentially uranyl structure like phosphate glasses, but with broader vibrational components (and hence less pronounced minima in the emission spectrum between such components) and the Stokes threshold is shifted to 502 nm (19900 cm^{-1}). The situation is more ambiguous in mixed borate-silicate glasses, where the emission spectrum shows quite broad maxima at 518 nm (19300) and 537 nm (18600) and a shoulder at 560 nm (17850 cm^{-1}). The latter glasses are not showing any absorption bands due to uranium(IV), but a rather weak, broad band appears at 500 nm. Ryan and Keder[197] reported $UO_2(NO_3)_4^{-2}$ and $UO_2(O_2CCH_3)_4^{-2}$ to have the fine-structure of the corresponding tris-complexes blurred out (like in borate glasses). It may be an intrinsic effect and not just a superposition of the spectra of non-equivalent uranyl groups.

An interesting case of sub-microscopic structure of glasses studied by luminescence is Eu(III) in various concentrations of phosphate and tungstate[198] showing that some of the samples are homogeneous and other are micro-crystallites dispersed in a vitreous matrix of differing composition. The studies have continued[199], in part also on glasses containing small uranyl concentrations.

Very effective *energy transfer* from the uranyl group to $^5D_0$ of Eu(III) is performed in phosphate glass[200]. Similar observations are reported[201] for energy transfer to the almost coinciding levels $^5S_2$ and $^5F_4$ (at 18500 cm^{-1}) of holmium(III). Increasing the weight concentration of holmium in barium borate glass from 0.1 to 3% increased the efficiency of energy transfer from 2.5 to 62.5% and, at the same time, τ decreased from 195 to 75 μs. It may be mentioned that even in aqueous solution[112, 202] europium(III) takes over energy from $*UO_2^{+2}$. A comparative study in phosphate glasses[25, 203] of energy transfer from the uranyl group to five trivalent lanthanides showed the probability to decrease along the series Nd(III) > Eu(III) > Pr(III) > Tm(III) > Dy(III) as a function of the distance between the Stokes threshold and the highest-lying *J*-level (below it) increasing from 1000 to 7000 cm^{-1}. For comparison, it may be mentioned that energy transfer from Nd(III) to $^2F_{5/2}$ of Yb(III) is quite effective[204] in germanate and tellurite glasses. Even the very short-lived luminescent state of bismuth(III) can transfer energy to Eu(III)[205, 206]. Energy transfer involving lanthanides has been reviewed recently[33, 187, 207]. Tanner and Vargenas[256] studied the pH dependence of energy transfer between $*UO_2^{+2}$ and Eu(III) in aqueous perchlorate solutions, and concluded in strong influence of oligomeric hydroxo complexes, probably containing OH^- bridges between uranyl and europium.

Such energy transfer may be important for *luminescent solar concentrators*[25, 86] trapping the major part of the light emitted by a fluorescent species (receiving solar radiation) by a series of total reflections between the parallel surfaces of a transparent, vitreous plate until it reaches the rim covered with a photovoltaic material (such as doped silicon). The purpose of this device is to decrease the amount of silicon needed by an order of magnitude, relative to the plate (perpendicular on the solar rays) being covered with silicon. This situation was realized[208] with a uranyl glass plate. However, the Stokes threshold around 20000 cm^{-1} is so high that only about a-fifth of the solar energy is absorbed (though flat-plate luminescent concentrators have the advantage that a large part of diffuse scattered day-light can be utilized) and the energy of the green photons emitted is so much higher than the energy gap of silicon that the excess energy has deleterious effects on the photoelectric yield. This difficulty can be circumvented[209] by energy transfer from the excited uranyl group to neodymium(III) or holmium(III), though ytterbium(III) would have been a better candidate[204]. The combination of the uranyl ion and lanthanides is not perfect[25], and further work[210] is in progress.

The energy transfer from $^*UO_2^{+2}$ in inorganic glasses to species such as Nd(III) or Eu(III) takes place as a purely spectroscopic excitation at 2.5 eV without taking into account the highly oxidizing character of the excited uranyl state discussed in Sect. 4. This does not prevent that phosphate glasses (and solutions in phosphoric acid) have a certain propensity[65] to form green uranium(IV) with the characteristic $5f^2$ absorption bands (which would be catastrophic for a luminescent solar concentrator). It is quite difficult to maintain definite oxidation states of transition-group elements in glasses[193] even when melting the glass in an oxidizing or a reducing atmosphere. A typical case is the formation of traces of CrO_4^{-2} in glasses[190] containing chromium(III) where small amounts of the mild reducing agent As_2O_3 can restore the pure Cr(III). We have experience with mixtures of UO_2^{+2} and manganese(II) in glasses producing high concentrations of U(IV).

Most chemists are rather unfamiliar with glasses in the sense that they expect a given element in a definite oxidation state to form a characteristic chromophore with a well-defined coordination number N and a locally fixed equilibrium geometry. Whereas many aqua ions are known to be a good approximation to octahedral $M(OH_2)_6^{+z}$ as known from the absorption spectra of V(II), V(III), Cr(III), Mn(II), Fe(III), Co(II), Co(III), Ni(II), Ru(II), Rh(III) and Ir(III), and from other physical properties of Al(III) and Ga(III), the lanthanide aqua ions are a much more intricate problem[5,6] and in general, the stereochemical preferences are very weakly pronounced for N above 6. It is unfortunate that text-books of geometry written by Euclid and subsequent authors classify the polyhedra (octahedron, dodecahedron, ...) according to their number of surfaces rather than the number of apices representing N for the chemist. Actually, for the chemist, the bonds to the adjacent atoms should rather be considered as vectors, having a definite direction, and a length corresponding to an internuclear distance R (or perhaps more appropriately, a strongly decreasing function of increasing R). The surfaces spanned by the closest-neighbour nuclei are not the slightest relevant for the chemist, in spite of a long-standing crystallographic tradition[211] of describing such polyhedra. It seems that trivalent lanthanides are very tolerant toward different symmetries and N for the vector needle-bunch[6,9] and the chemical bonding has hardly any angular dependence. By the same token, it is quite likely that Li(I), Na(I), Ca(II) and Th(IV) aqua ions in solution represent a great dispersion of N and shape of vector bunches on an instantaneous picture.

Brawer and Weber[212] performed 100 repeated Monte Carlo calculations on a glass represented by the cubic unit cell (with Eu(III) at the centre) containing $Be_{65}EuF_{133}$ (BeF_2 is very similar to SiO_2) and for Eu–F distances

	$N =$ 5	6	7	8	9	
$R < 2.7$ Å	0.09	0.29	0.47	0.15	0.00	(14)
$R < 3.0$ Å	0.00	0.12	0.34	0.48	0.06	

whereas Be–F distances determine $N = 3$ and 4 (like in cristobalite-type BeF_2). If additional cations such as Na^+, Rb^+ or Ca^{+2} in $Be_{76}Na_{74}EuF_{229}$ ... are added to the glass. N for Be(II) increases to 5 for a-third to a-half, and N increases also in direction of 9 and 10 for Eu(III). There is no doubt that uranyl glasses present additional problems of this kind because of the anisotropic bonding[2, 61] in the equatorial plane. Brawer and Weber[213] performed analogous studies of Nd(III) in fluoroberyllate glasses.

10 Neptunyl, Plutonyl and Americyl Ions

Since the uranyl ion is so obviously its own category, it is very interesting to compare with the analogous species formed by transuranium elements. M = Np, Pu and Am form all three MO_2^{+2} and MO_2^+ which are, by no means, the most stable oxidation states of their elements, and which tend toward reduction by the radiochemical products concomitant with the high specific radioactivity of the isotopes normally studied of plutonium and americium (whereas e.g. ^{244}Pu with the half-life 82 million years would not present this problem). Contrary to some reports in literature, it does not seem that curium (and the subsequent elements) form such dioxo complexes.

It was early pointed out[23, 214] that only the MO_2^{+2} show electron transfer bands in the visible, whereas the weak, but narrow, bands of MO_2^+ seem all to be due to internal transitions in the partly filled 5f shell. In this sense, even Am(V) seems less oxidizing than U(VI) though it is surprising how small a red shift (the first maxima move from 20600 cm^{-1} in UO_2^{+2} to 18000 in NpO_2^{+2}, 17000 or 19200 cm^{-1} in PuO_2^{+2} and about 13000 cm^{-1} in AmO_2^{+2}) there is from $5f^0$ U(VI) to $5f^3$ Am(VI). In one way, this corroborates the chemical observation[3, 18] that plutonium and americium belong to the group of elements (like chromium, manganese, ruthenium, xenon and osmium) which are more able to attain high oxidation states with oxide than with fluoride ligands. We are now discussing the individual MO_2^{+2} and MO_2^+ arranged according to increasing number of 5f electrons in the groundstate.

The pink NpO_2^{+2} has a very sharp band measured in 1 molar $D_3O^+ClO_4^-$ in deuterium oxide[215] at 8180 cm^{-1} ($\varepsilon = 45$) which is likely to represent the transition from the lowest Ω level of $^2F_{5/2}$ (we gave arguments in Sect. 3 that it is $\Omega = 5/2$ in agreement[78] with one of two determinations by paramagnetic resonance) to either $\Omega = 7/2$ or 5/2 (or both) of $^2F_{7/2}$. Originally, there was an expectation of very large π- and σ-anti-bonding effects, whereas the work on $NpO_2Cl_4^{-2}$ syncrystallized in $Cs_2UO_2Cl_4$[79] represents the opposite extreme, the $^2F_{7/2}$ sub-levels all supposed to be close to 6880 cm^{-1}, and the first electron transfer band occurring already at 13300 cm^{-1}.

In the original spectra of NpO_2^{+2} measured by Sjoblom and Hindman[216] two bands are clearly seen at 18000 ($\varepsilon = 6.8$) and 21000 cm^{-1} ($\varepsilon = 6.4$). It is almost inconceivable that the electron transfer bands of the uranyl ion do not have analogs at slightly lower wave-numbers in the neptunyl(VI) ion. However, it is less certain whether the weak bands of NpO_2^{+2} at 16200 ($\varepsilon = 2.1$) and 16900 cm^{-1} ($\varepsilon = 3.2$) correspond to the first, weak electron transfer band of UO_2^{+2} at 20600 cm^{-1} or rather are "ligand field" transitions in $5f^1$. The neptunyl chloro complex[217] extracted by a 10% solution of $N(C_8H_{17})_3$ in xylene from aqueous hydrochloric acid has a group of three weak bands starting at

13 000 cm^{-1} and a group of four stronger bands starting at 15 100 cm^{-1}. When same experiment is performed with uranyl chloride, the first sharp band occurs at 20 200 cm^{-1}. Ryan[218] obtained rather similar spectra on ion-exchange resins of a species argued to be $[(H_3O)(H_2O)_3(MO_2Cl_4)_2]^{-3}$. With M = Np, the rather structured group at 6800 cm^{-1} is more intense than the broad band at 8000 cm^{-1}, in contrast to aqueous NpO_2^{+2} [215] where the band at 6750 cm^{-1} is 9 times weaker than the peak at 8180 cm^{-1}. Again, three weak bands start at 13 200 cm^{-1} and the four stronger bands at 15 600 cm^{-1}. Disregarding pronounced effects of hydrogen bonding[218] this spectrum is very similar to $NpO_2Cl_4^{-2}$ syncrystallized in $Cs_2UO_2Cl_4$[79] with electronic origins at 13 277, 15 426, 17 478 and 19 358 cm^{-1}. There is no doubt that the two band maxima of aqueous NpO_2^{+2} [215, 216] at 18 000 and 21 000 cm^{-1} correspond to the two last origins in the cooled crystal. Even the origin at 13 277 cm^{-1} may be due to electron transfer, but a comparison between $UO_2Cl_4^{-2}$ and other uranyl salts in Table 2 makes it then probable that the electron is transferred from a M.O. mainly localized on the oxo (and not the chloro) ligands. However, an alternative is that $^2F_{7/2}$ is situated 7700 cm^{-1} above $^2F_{5/2}$ and that $5f\varphi$ and $5f\delta$ have vanishing anti-bonding effect (it is not absolutely excluded[69, 77] that the equatorial ligands push $5f\varphi$ slightly above $5f\delta$) whereas $5f\pi$ is anti-bonding to the extent 7000 (Case A) or 10 500 cm^{-1} (Case B) and $5f\sigma$ to the extent 8400 (A) or 21 000 cm^{-1} (B). In the unit 1000 cm^{-1}, the eigen-values[23] are:

Ω =	5/2	3/2	7/2 & 5/2	1/2	3/2	1/2	
Case A:	0	1.2	7.7	7.0	13.5	16.1	(15)
Case B:	0	1.4	7.7	12.6	16.8	26.6	

Obviously, other alternatives are conceivable, but the main choice is whether the two bands in the red correspond to the two highest eigen-values, Ω = 3/2 below 1/2 (case A) or the other way round, the third- and the next-highest eigen-value (case B) with the highest Ω = 1/2 lost at very high wave-numbers. The various M.O. calculations summarized in Table 1 are closer related to case B. In Eq. (15), case A represents the ultimate occasion to keep the highest $5f^1$ energy level below the first electron transfer band. As soon this is no longer the case, the separation between the weak band at 6750 and the strong at 8180 cm^{-1} in the aqua ion[215] is rather connected with Ω = 7/2 and 5/2 not being perfectly degenerate.

It would be of great interest to obtain information about the isoelectronic UO_2^+. Unfortunately, much less is known about this species than about UX_6^- [67, 179], UOX_5^{-2} [179], UO_5F^{-6} [186] and UO_6^{-7} [178] where X are halogens. For our purposes, the most important conclusion is the "ligand field" treatment of UX_6^- [219, 220] where the three (simultaneously σ- and π-anti-bonding) t_{1u} 5f-like orbitals are situated 11 300 cm^{-1} above the non-bonding a_{2u} orbital in UF_6^- and 6400 cm^{-1} in UCl_6^-, whereas these figures are 4500 and 2900 cm^{-1} for the three π-anti-bonding t_{2u} orbitals relative to a_{2u} (cf. Table 3). One of several difficulties is that Newton and Baker[131] found an adduct of UO_2^+ and UO_2^{+2} having a definite band at 737 nm (13 550 cm^{-1}). In their review on U(V) chemistry and spectra[221] Selbin and Ortego ascribed this band to U(V) alone, but Ekstrom[222] confirmed the band (with ε = 24.3) as belonging to $U_2O_4^{+3}$. This dimer has also a broad band in the blue, with the maximum close to 410 nm having ε = 28, probably belonging to the U(VI) fragment. The quilibrium constant is 15 ± 5 M^{-1}. UO_2^+ disproportionates rapidly to U(IV) and U(VI), but with the lowest rate at pH close to 2. Bell, Friedman and

Billings[223] studied the absorption spectrum of a reddish orange aqueous perchlorate solution of UO_2^+ showing a rather indistinct slope between 500 and 650 nm, and bands at 738 (ε = 12) and 940 nm (ε = 5.7). Cohen[224] studied UO_2^+ in a saturated solution of calcium chloride and found a strong peak at 1510 nm (6660 cm^{-1}) probably of the same origin as at 8180 cm^{-1} in NpO_2^+. U(V) carbonate complexes have been studied with various experimental techniques[114, 224].

We discussed above[127–129] the adducts of NpO_2^+ with Cr(III), Rh(III) and other trivalent aqua ions, and Th(IV), and[225] also between NpO_2^+ and UO_2^{+2} where the formation constant K = 0.7 M^{-1} for the adduct $NpUO_4^{+3}$ was redetermined by Madic et al.[132] to be K = 3.7 M^{-1}. These authors evaluated also the sharp band at 992 nm with $\varepsilon \sim 260$, and a weaker band at 840 nm. Essentially the same band position 992 nm ($\varepsilon \sim 230$) was found for the adduct between NpO_2^+ and NpO_2^{+2} having K = 3.0 M^{-1}.

The aqueous solution of isolated NpO_2^+ has a large number of sharp bands[216] besides the high peak at 980 nm. In view of the significant "ligand field" effects in NpO_2^{+2} and UX_6^- one cannot hope for the 13 *J*-levels of $5f^2$ in spherical symmetry to be readily recognized in Np(V). The situation is much more favourable in U(IV) with "ligand field" effects about 4 times smaller[219, 220] than in UX_6^-. Further on, gaseous U^{+4} was recently analyzed[160] and the nephelauxetic effect in U(IV) aqua ions evaluated[226] and shown to be quite moderate, slightly more pronounced than in $4f^2$ praseodymium(III) (where it is strongest among the trivalent lanthanides) and slightly weaker than in $3d^5$ manganese(II) (showing the weakest nephelauxetic effect among the d-group central atoms) with a given set of ligating atoms. Though N = 8, 9, 10, ... (and in several organometallic complexes 16) are frequent for U(IV), the octahedral UCl_6^{-2} and UBr_6^{-2} have been studied thoroughly[227]. Among the 91 states of $5f^2$, the point-group O_h induces 40 sub-levels Γ_J and 21 of those have been firmly identified. Further on, the very high-lying thirteenth *J*-level $^1S_0$ known[214, 228] at 40800 cm^{-1} in uranium(IV) aqua ions, can now be compared with the other *J*-levels of U^{+4}. When comparing the (excited) configuration $[86]5f^2$ in Th^{+2} with U^{+4} the situation is less simple than predicted by Gruen[126] and the percentage increase for differing *J*-level positions is between 21 and 51, whereas there is a much more regular increase of *J*-level distances[31] from Pr^{+2} to the isoelectronic Nd(III) by multiplication by a factor close to 1.23 (and it is likely that this factor is 1.26 for $[54]4f^3$ of the Nd^{+3} not yet analyzed).

The question of much stronger "ligand field" effects in M(V) than in M(IV) of transthorium elements does not seem to be connected with a specific property of the 5f group, but rather with the higher oxidation state. Thus, already in the $5f^3$ U(III) the "ligand field" effects are only twice those known from the 4f group, as can be seen most clearly[229, 230] from a comparison between $U_xLa_{1-x}Cl_3$ and $Nd_xLa_{1-x}Cl_3$.

Hence, there are convincing reasons to expect strong chemical effects on the $5f^2$ system Pu(VI) known from PuO_2^{+2} and PuF_6. Unfortunately, the electron transfer bands are lower (though broader) in the visible than the $5f^2$ internal transitions. Two candidates for the lowest wave-number of such an electron transfer band in PuO_2^{+2}[253] are situated at 17000 or 19200 cm^{-1}. The spectroscopic difference that MO_2^+ does not seem to have electron transfer spectra in the visible is accompanied by a maximum chemical stability of NpO_2^+. However, the rate of ^{18}O exchange at 23 °C in 1 M perchloric acid is 0.31 s^{-1} in NpO_2^+ (PuO_2^+ is much slower, UO_2^+ much more rapid) but below $6 \cdot 10^{-7}$ s^{-1} in NpO_2^{+2}[231]. Much of the reported chemistry of transuranium elements is influenced by redox reactions, due to products of the intense radioactivity. Thus, 1 g of the uranium isotope mixture extracted from minerals emit as many α-particles per minute as 1 mg ^{237}Np or 0.01 mg ^{239}Pu or $2 \cdot 10^{-7}$ g ^{241}Am.

It is not perfectly clear where the first electron transfer band of AmO_2^{+2} is situated. Two narrow 5 f^3 transitions occur at 10080 cm^{-1} ($\varepsilon = 72$) and at 15000 cm^{-1} ($\varepsilon = 20$) but the broad electron transfer bands are difficult to ascertain[232]. However, more recent measurements[233, 234] show a band at 13000 cm^{-1} ($\varepsilon = 10$) likely to be the first uranyl-like band. On the other hand, Am(VI) solutions in strong sodium carbonate or bicarbonate treated with ozone go red-brown[235] and develop a strong absorption edge in the yellow (ε crosses 140 to 610 nm) and 0.1 molar $Na_4P_2O_7$, an edge passes $\varepsilon = 130$ at 530 nm. These electron transfer bands are due to ligands in the equatorial plane, like red $UO_2I_4^{-2}$ or uranyl complexes of sulphur-containing ligands. However, the main reason to suspect the bands[233] at 13200 and 13700 cm^{-1} for being uranyl-like is that the sharp peak at 15080 cm^{-1} is far narrower.

None of the transuranium MO_2^+ and MO_2^{+2} have been reported to fluoresce. The easiest explanation is the cross-over (at highly different nuclear positions) of the potential surfaces[58] of the electron transfer state and the groundstate. However, we have to realize that the Stokes shift of UO_2^{+2} and UF_6 is not excessive, compared with bismuth(III) in solids[41, 236] and especially in cases of Pb(II) or Bi(III) co-existing[60] with MoO_4^{-2}, WO_4^{-2}, WO_6^{-6} and VO_4^{-3}. In their review on energy transfer[207] Powell and Blasse point out the enormous Stokes shift of the first excited state $^3T_{1g}$ of $Co(CN)_6^{-3}$ having one d-like electron excited from the lower to the higher sub-shell. The question of luminescence of electron transfer states of d-group complexes was recently put in a new light by Flint and Paulusz[237] showing that $Cs_2(IrCl_6)_{0.001}(SnCl_6)_{0.999}$ at 5 K emits 16 broad features (roughly equidistant 300 cm^{-1}) between 12000 and 16000 cm^{-1} belonging either to the electron transfer $(\pi)t_{1g} \rightarrow (5d)t_{2g}$ giving the first maximum at 17000 cm^{-1}, or to $(\pi + \sigma)t_{1u} \rightarrow (5d)t_{2g}$ with very large changes of nuclear positions. Anyhow, the maximum intensity occurs at the 7th or 8th vibrational components. These bands shift in organic solvents[238] and at least the first electron transfer band moves to lower wave-numbers at high pressure, indicating the minimum of the potential surface to occur at lower internuclear distances than in the groundstate. Like in manganese(II) complexes, this case enhances the yield of luminescence[25] because of less pushing of the surrounding atoms.

We would like to mention the fascinating *triboluminescence*[239] of uranyl nitrate, where the emitted spectrum looks like photoluminescence, whereas several other triboluminescent materials crushed in nitrogen or neon give the characteristic spectra of electric discharges in the surrounding gas.

Finally, we would like to draw attention to the thorough review of 5 f-group absorption spectra by Ryan[240].

11 Present Status of an Old Coral Reef of Problems

The uranyl ion has a totally symmetric groundstate[28] with $\Omega = 0^+$. Hence, it is not open to the same kind of controversy as was quite vigorous around 1955[17, 241] about whether uranium belongs to the 5 f group or is a higher homolog of chromium, molybdenum and tungsten, and hence is a 6 d element. It turned out that all non-metallic compounds of the transthorium elements have ($Z - 86 - z$) electrons in the 5f shell, where z is the oxidation state[31]. On the other hand, it has also turned out that there is no necessary relation between the presence of f electrons and a strong propensity toward almost invariant trivalency. At most, there is a trend in the 4f group[9, 33] and in the elements starting with americium[31, 242] to have an invariant oxidation state (corresponding to the huge difference between the ionization energy and electron affinity of the partly filled f shell, much like when a closed-shell electron configuration is crossed as a function of increasing oxidation state) but it is by no means obvious that it should be + 3.

There was a time when the (frequently fluorescent) first excited state was thought to be an intrinsic characteristic of the triatomic uranyl group, much like the blue colour of ozone or S_3^- now known to produce the colour of ultramarine (lapis lazuli). When it was realized[23, 53] that it is the lowest-lying member of a complicated manifold of electron transfer states belonging to an electron configuration with one 5f electron and a hole in a M.O. mainly delocalized on the two oxygen atoms, it became also apparent that there is no proportion between the chemically oxidizing character and the facts that both the uranyl ion and chromate have their first electron transfer band in the blue, at 30000 and 23500 cm^{-1} lower energy than WO_4^{-2} in solution. This discrepancy is attenuated by the anisotropic properties[55, 61] of the uranyl and neptunyl complexes of strongly reducing ligands in the equatorial plane, attaining wave-numbers of the electron transfer bands agreeing somewhat more closely with chemical expectations. Thus, the fox-red colour of uranyl complexes of sulphur-containing ligands corresponds to the first electron transfer band at a wave-number situated between 21500 cm^{-1} of red MoS_4^{-2} and 25500 cm^{-1} of orange WS_4^{-2} [18].

Nevertheless, as we have reviewed, there are enigmatic aspects of the electronic states of both UO_2^{+2} and of UF_6. As pointed out by Dr. Ruth McDiarmid[142, 152] there is an enormous shift from the weak band of UF_6 at 25800 cm^{-1} to the first band of WF_6 at about 58000 cm^{-1} whereas the chemical reduction to UF_6^- is not much more eager than to WF_6^-. As seen from Tables 1 and 3, the recent relativistic M.O. calculations have not clarified these problems completely. As soon two or more calculations have been published they show a large dispersion in the predictions of the various details of interest to us.

For the chemist, the long-lived fluorescent state of the uranyl ion has the highly specific characteristic that (besides interesting reactions of hydrogen atom abstraction from organic molecules) the formation of exciplexes studied by Professor Marcantonatos and the reactions of the excited state with specific cations (such as silver(I), thallium(I) and manganese(II)) and anions (showing formation constants entirely different from the uranyl groundstate, and more appropriate for a palladium-like element) show how our knowledge is sporadic, and how inorganic chemistry still seems to reveal endless supplies of surprises.

Acknowledgments. We would like to thank Professor Minas D. Marcantonatos for many helpful and interesting discussions.

We also thank Professor Colin D. Flint, Birkbeck College, University of London, for a generous supply of preprints.

We are grateful to the Swiss National Science Foundation (grant no. 2.841-080) assisting this long-distance collaboration by allowing frequent periods of work in Geneva.

12 References

1. Jørgensen, C. K.: Rev. chim. min. (Paris) *14,* 127 (1977)
2. Burrows, H. D., Kemp, T. J.: Chem. Soc. Rev. (London) *3,* 139 (1974)
3. Jørgensen, C. K.: Inorganic Complexes. London, Academic Press 1963
4. Jørgensen, C. K.: Naturwissenschaften *67,* 188 (1980)
5. Williams, R. J. P.: this volume
6. Jørgensen, C. K., Reisfeld, R.: Topics Current Chem. *100,* 127 (1982)
7. Jørgensen, C. K.: Topics Current Chem. *56,* 1 (1975)
8. Jørgensen, C. K.: Chimia *23,* 292 (1969)
9. Jørgensen, C. K.: Handbook on the Physics and Chemistry of Rare Earths (K. Gschneidner, L. Eyring, eds.) *3,* p. 111, North-Holland, Amsterdam 1979
10. Jørgensen, C. K.: Comments Inorg. Chem. *1,* 123 (1981)
11. Block, B. P., Bailar, J. C.: J. Am. Chem. Soc. *73,* 4722 (1951)
12. Skibsted, L. H., Bjerrum, J.: Acta Chem. Scand. *A 28,* 740 and 764 (1974)
13. Jørgensen, C. K.: Acta Chem. Scand. *11,* 73 (1957)
14. Zeltmann, A. H., Morgan, L. O.: Inorg. Chem. *10,* 2739 (1971)
15. Jørgensen, C. K.: Structure and Bonding *6,* 94 (1969)
16. Weeks, M. E.: The Discovery of the Elements (7. ed). Easton, Penn., J. Chem. Educ. Publishers 1968
17. Jørgensen, C. K.: J. chim. physique *76,* 630 (1979)
18. Müller, A., Diemann, E., Jørgensen, C. K.: Structure and Bonding *14,* 23 (1973)
19. Kanzelmeyer, J. H., Ryan, J., Freund, H.: J. Am. Chem. Soc. *78,* 3020 (1956)
20. Stiefel, E. I.: Progress Inorg. Chem. *26,* 1 (1979)
21. Tatsumi, K., Hoffmann, R.: Inorg. Chem. *19,* 2656 (1980)
22. Rabinowitch, E., Belford, R. L.: Spectroscopy and Photochemistry of Uranyl Compounds. Oxford, Pergamon 1964
23. Jørgensen, C. K.: Acta Chem. Scand. *11,* 166 (1957)
24. Bell, J. T., Biggers, R. E.: J. Mol. Spectr. *18,* 247 (1965)
25. Reisfeld, R., Jørgensen, C. K.: Structure and Bonding *49,* 1 (1982)
26. Pais, A.: Rev. Mod. Phys. *49,* 925 (1977)
27. Glashow, S. L.: Rev. Mod. Phys. *52,* 539 (1980)
28. Jørgensen, C. K.: Structure and Bonding *34,* 19 (1978) and *43,* 1 (1981)
29. Jørgensen, C. K.: Nature *292,* 41 (1981)
30. Jørgensen, C. K.: Orbitals in Atoms and Molecules. London, Academic Press 1962
31. Jørgensen, C. K.: Oxidation Numbers and Oxidation States. Berlin – Heidelberg – New York, Springer-Verlag 1969
32. Jørgensen, C. K.: Modern Aspects of Ligand Field Theory. Amsterdam, North-Holland 1971
33. Reisfeld, R., Jørgensen, C. K.: Lasers and Excited States of Rare Earths. Berlin – Heidelberg – New York, Springer-Verlag 1977
34. Jørgensen, C. K.: Int. Rev. Phys. Chem. *1,* 225 (1981)
35. Jørgensen, C. K.: Adv. Chem. Phys. *5,* 33 (1963)
36. Jørgensen, C. K.: Progress Inorg. Chem. *12,* 101 (1970)
37. Jørgensen, C. K.: Mol. Phys. *2,* 309 (1959)
38. Jørgensen, C. K.: Solid State Phys. *13,* 375 (1962)
39. Jørgensen, C. K.: Halogen Chemistry (V. Gutmann, ed.) *1,* p. 265, London, Academic Press 1967
40. Walton, R. A., Matthews, R. W., Jørgensen, C. K.: Inorg. Chim. Acta *1,* 355 (1967)
41. Boulon, G., Jørgensen, C. K., Reisfeld, R.: Chem. Phys. Letters *75,* 24 (1980)
42. Jørgensen, C. K.: Israel J. Chem. *19,* 174 (1980)
43. Ryan, J. L., Jørgensen, C. K.: J. Phys. Chem. *70,* 2845 (1966)
44. Ryan, J. L.: Inorg. Chem. *8,* 2053 (1969)
45. Linhard, M., Weigel, M.: Z. anorg. Chem. *266,* 49 (1951)
46. Jørgensen, C. K.: Structure and Bonding *22,* 49 (1975)
47. Jørgensen, C. K., Berthou, H.: Mat. fys. Medd. Dan. Vid. Selskab *38,* no. 15 (1972)
48. Jørgensen, C. K.: Structure and Bonding *24,* 1 (1975)

49. Turner, D. W., Baker, C., Baker, A. D., Brundle, C. R.: Molecular Photoelectron Spectroscopy. London, Wiley-Interscience 1970
50. Rabalais, J. W.: Principles of Ultraviolet Photoelectron Spectroscopy. New York, Wiley-Interscience 1977
51. Furlani, C., Cauletti, C.: Structure and Bonding *35,* 119 (1978)
52. Jørgensen, C. K.: Adv. Quantum Chem. *11,* 51 (1978)
53. Jørgensen, C. K., Reisfeld, R.: Chem. Phys. Letters *35,* 441 (1975)
54. Jørgensen, C. K.: Mol. Phys. *5,* 271 (1962)
55. Ryan, J. L., Jørgensen, C. K.: Mol. Phys. *7,* 17 (1963)
56. Taylor, J. C., Wilson, P. W.: Acta Cryst. *B30,* 1481 (1974)
57. Blasse, G., Bril, A.: J. Chem. Phys. *45,* 3327 (1966)
58. Blasse, G.: Structure and Bonding *26,* 43 (1976)
59. Duffy, J. A.: Structure and Bonding *32,* 147 (1977)
60. Blasse, G.: Structure and Bonding *42,* 1 (1980)
61. Jørgensen, C. K.: Proceed. Symposium Tihany 1964, p. 11. Budapest, Akadémia Kiadó 1965
62. Herzberg, G.: Electronic Spectra and Electronic Structure of Polyatomic Molecules. Princeton, Van Nostrand 1966
63. Kaufman, V., Radziemski, L. J.: J. Opt. Soc. Am. *66,* 599 (1976)
64. Jørgensen, C. K.: J. Inorg. Nucl. Chem. *1,* 301 (1955)
65. Lieblich-Sofer, N., Reisfeld, R., Jørgensen, C. K.: Inorg. Chim. Acta *30,* 259 (1978)
66. Jørgensen, C. K.: J. Luminescence *18,* 63 (1979)
67. Reisfeld, M. J., Crosby, G. A.: Inorg. Chem. *4,* 65 (1965)
68. Yang, C. Y., Johnson, K. H., Horsley, J. A.: J. Chem. Phys. *68,* 1001 (1978)
69. Walch, P. F., Ellis, D. E.: J. Chem. Phys. *65,* 2387 (1976)
70. Boring, M. Wood, J. H.: J. Chem. Phys. *71,* 392 (1979)
71. Veal, B. W., Lam, D. J., Carnall, W. T., Hoekstra, H. R.: Phys. Rev. *B12,* 5651 (1975)
72. Jørgensen, C. K.: Structure and Bonding *30,* 141 (1976)
73. Denning, R. G., Snellgrove, T. R., Woodwark, D. R.: Mol. Phys. *32,* 419 (1976)
74. DiSipio, L., Tondello, E., Pelizzi, G., Ingletto, G., Montenero, A.: Inorg. Chim. Acta *22,* 23 (1977)
75. Flint, C. D., Tanner, P. A.: J. C. S. Faraday Trans. II *74,* 2210 (1978)
76. Flint, C. D., Tanner, P. A.: J. Luminescence *18,* 69 (1979)
77. Denning, R. G., Foster, D. N. P., Snellgrove, T. R., Woodwark, D. R.: Mol. Phys. *37,* 1089 and 1109 (1979)
78. Leung, A. F., Wong, E. Y.: Phys. Rev. *180,* 380 (1969)
79. Stafsudd, O. M., Leung, A. F., Wong, E. Y.: Phys. Rev. *180,* 339 (1969)
80. Görller-Walrand, C., Vanquickenborne, L. G.: J. Chem. Phys. *54,* 4178 (1971)
81. Görller-Walrand, C., DeJaegere, S.: Spectrochim. Acta *A28,* 257 (1972)
82. Görller-Walrand, C., DeJaegere, S.: J. chim. physique *69,* 726 (1972)
83. Flint, C. D., Tanner, P. A.: Mol. Phys. *43,* 933 (1981)
84. Blasse, G.: Chem. Phys. Letters *63,* 441 (1979)
85. Balzani, V., Bolletta, F., Gandolfi, M. T., Maestri, M.: Topics Current Chem. *75,* 1 (1978)
86. Jørgensen, C. K.: Photochemical Conversion and Storage of Solar Energy (J. S. Connolly, ed.) p. 79. New York, Academic Press 1981
87. Balzani, V., Scandola, F.: ibid., p. 97
88. Schumacher, E.: Chimia *32,* 193 (1978)
89. Calzaferri, G.: Chimia *32,* 241 (1978) and *35,* 209 (1981)
90. Kiwi, J., Kalyanasundaram, K., Grätzel, M.: Structure and Bonding *49,* 37 (1982)
91. Tributsch, H.: Structure and Bonding *49,* 127 (1982)
92. Kraus, K. A., Nelson, F.: J. Am. Chem. Soc. *71,* 2517 (1949)
93. Marcantonatos, M. D.: J. C. S. Faraday Trans. I *75,* 2252 (1979)
94. Claudel, B., Sautereau, H.: Spectrochim. Acta *A29,* 1687 (1973)
95. Sostero, S., Traverso, O., Bartocci, C., DiBernardo, P., Magon, L., Carassiti, V.: Inorg. Chim. Acta *19,* 229 (1976)
96. Claudel, B., Fève, M., Puaux, J. P., Sautereau, H.: J. Photochem. *7,* 113 (1977) and *8,* 117 (1978)
97. Kropp, J. L., Windsor, M. W.: J. Chem. Phys. *45,* 761 (1966)

98. Haas, Y., Stein, G.: J. Chem. Phys. *75,* 3668 and 3677 (1971)
99. Jørgensen, C. K.: Chimia *34,* 381 (1980)
100. Marcantonatos, M. D.: J. C. S. Faraday Trans. I *75,* 2273 (1979)
101. Deschaux, M., Marcantonatos, M. D.: Chem. Phys. Letters *63,* 283 (1979)
102. Marcantonatos, M. D.: J. C. S. Faraday Trans. I *76,* 1093 (1980)
103. Moriyasu, M., Yokoyama, Y., Ikeda, S.: J. Inorg. Nucl. Chem. *39,* 2199 and 2211 (1977)
104. Pant, D. D., Pant, H. C.: Indian J. Pure Appl. Phys. *6,* 219 (1968)
105. Joshi, G. C.: Indian J. Pure Appl. Phys. *14,* 180 (1976)
106. Ahrland, S.: Structure and Bonding *5,* 118 (1968) and *15,* 167 (1973)
107. Marcantonatos, M. D., Deschaux, M., Celardin, F.: Chem. Phys. Letters *69,* 144 (1980)
108. Marcantonatos, M. D., Deschaux, M., Celardin, F., Levental, M.: Chem. Phys. Letters *65,* 316 (1979)
109. Jørgensen, C. K., Parthasarathy, V.: Acta Chem. Scand. *A 32,* 957 (1978)
110. Deschaux, M., Marcantonatos, M. D.: J. Inorg. Nucl. Chem. *43,* 361 (1981)
111. Giguère, P. A.: Chem. Phys. Letters *41,* 598 (1976) and J. Chem. Educ. *56,* 571 (1979)
112. Matsushima, R., Fujimori, H., Sakuraba, S.: J. C. S. Faraday Trans. I *70,* 1702 (1974)
113. Ciavatta, L., Ferri, D., Grimaldi, M., Palombari, R., Salvatore, F.: J. Inorg. Nucl. Chem. *41,* 1175 (1979)
114. Wester, D. W., Sullivan, J. C.: Inorg. Chem. *19,* 2838 (1980)
115. Burrows, H. D., Pedrosa de Jesus, J. D.: J. Photochem. *5,* 265 (1976)
116. Döller, E., Förster, T.: Z. physik. Chem. *34,* 132 (1962)
117. Marcantonatos, M. D.: Inorg. Chim. Acta *26,* 41 (1978)
118. Dunsmore, H. S., Hietanen, S., Sillén, L. G.: Acta Chem. Scand. *17,* 2644 (1963)
119. Burrows, H. D., Formosinho, S. J., Miguel, M. da G., Pinto Coelho, F.: J. C. S. Faraday Trans. I *72,* 163 (1976)
120. Jørgensen, C. K.: Acta Chem. Scand. *8,* 1502 (1954)
121. Jørgensen, C. K.: Mol. Phys. *4,* 235 (1961)
122. Jørgensen, C. K.: Acta Chem. Scand. *17,* 1034 and 1043 (1963)
123. Marcantonatos, M. D.: Deschaux, M.: Chem. Phys. Letters *76,* 359 (1980)
124. Marcantonatos, M. D., Deschaux, M.: Chem. Phys. Letters *80,* 327 (1981)
125. Deschaux, M., Marcantonatos, M. D.: J. Inorg. Nucl. Chem. *43,* 2915 (1981)
126. Gruen, D. M.: J. Chem. Phys. *20,* 1818 (1952)
127. Sullivan, J. C.: J. Am. Chem. Soc. *84,* 4256 (1962)
128. Sullivan, J. C.: Inorg. Chem. *3,* 315 (1964)
129. Murmann, R. K., Sullivan, J. C.: Inorg. Chem. *6,* 892 (1967)
130. Newton, T. W., Baker, F. B.: Inorg. Chem. *1,* 368 (1962)
131. Newton, T. W., Baker, F. B.: Inorg. Chem. *4,* 1166 (1965)
132. Madic, C., Guillaume, B., Morisseau, J. C., Moulin, J. P.: J. Inorg. Nucl. Chem. *41,* 1027 (1979)
133. Allen, D. M., Burrows, H. D., Cox, A., Hill, R. J., Kemp, T. J., Stone, T. J.: J. C. S. Chem. Comm. 59 (1973)
134. Sergeeva, G., Chibisov, A., Levshin, L., Karyakin, A.: J. C. S. Chem. Comm. 159 (1974)
135. Hill, R. J., Kemp, T. J., Allen, D. M., Cox, A.: J. C. S. Faraday Trans. I *70,* 847 (1974)
136. Reinen, D., Friebel, C.: Structure and Bonding *37,* 1 (1979)
137. Flint, C. D.: Chemistry in Britain *14,* 223 (1978)
138. Wong, D. P., Wong, A. Y. C., Wong, E. Y.: J. Chem. Phys. *56,* 2838 (1972)
139. Steindler, M. J., Gerding, T. J.: Spectrochim. Acta *22,* 1197 (1966)
140. Steindler, M. J., Gunther, W. H.: Spectrochim. Acta *20,* 1319 (1964)
141. Jørgensen, C. K.: Acta Chem. Scand. *17,* 251 (1963)
142. McDiarmid, R.: J. Chem. Phys. *65,* 168 (1976)
143. Lewis, W. B., Asprey, L. B., Jones, L. H., Mc Dowell, R. S., Rabideau, S. W., Zeltmann, A. H., Paine, R. T.: J. Chem. Phys. *65,* 2707 (1976)
144. Grzybowski, J. M., Andrews, L.: J. Chem. Phys. *68,* 4540 (1978)
145. Bagnall, K. W., Du Preez, J. G. H., Gellatly, B. J., Holloway, J. H.: J. C. S. Dalton 1963 (1975)
146. Allen, G. C., Warren, K. D.: Structure and Bonding *9,* 49 (1971) and *19,* 105 (1974)
147. Case, D. A., Yang, C. Y.: J. Chem. Phys. *72,* 3443 (1980)

148. Koelling, D. D., Ellis, D. E., Bartlett, R. J.: J. Chem. Phys. *65,* 3331 (1976)
149. Hay, P. J., Wadt, W. R., Kahn, L. R., Raffenetti, R. C., Phillips, D. C.: J. Chem. Phys. *71,* 1767 (1979)
150. Karlsson, L., Mattson, L., Jadrny, R., Bergmark, T., Siegbahn, K.: Physica Scripta (Stockholm) *14,* 230 (1976)
151. Rösch, N., Smith, V. H., Whangbo, H. H.: J. Am. Chem. Soc. *96,* 5984 (1974)
152. McDiarmid, R.: Chem. Phys. Letters *76,* 300 (1980)
153. Pyykkö, P.: Adv. Quantum Chem. *11,* 353 (1978)
154. Pyykkö, P., Desclaux, J. P.: Accounts Chem. Res. *12,* 276 (1979)
155. Pitzer, K. S.: Accounts Chem. Res. *12,* 271 (1979)
156. Fricke, B.: Structure and Bonding *21,* 89 (1975)
157. Jørgensen, C. K.: Lanthanides and Elements from Thorium to 184. London, Academic Press, in preparation
158. Pyper, N. C., Grant, I. P.: J. C. S. Faraday Trans. II *74,* 1885 (1978)
159. Jørgensen, C. K.: Angew. Chem. *85,* 1 (1973) and Int. Ed. *12,* 12 (1973)
160. Wyart, J. F., Kaufman, V., Sugar, J.: Physica Scripta (Stockholm) *22,* 389 (1980)
161. Ferreira, R.: Structure and Bonding *31,* 1 (1976)
162. De Mello, P. C., Hehenberger, M., Larsson, S., Zerner, M.: J. Am. Chem. Soc. *102,* 1278 (1980)
163. Thornton, G., Rösch, N., Edelstein, N.: Inorg. Chem. *19,* 1304 (1980)
164. Thornton, G., Edelstein, N., Rösch, N., Egdell, R. G., Woodwark, D. R.: J. Chem. Phys. *70,* 5218 (1979)
165. Srivastava, S. K., Cartwright, D. C., Trajmar, S., Chutjilan, A., Williams, W. X.: J. Chem. Phys. *65,* 208 (1976)
166. Rianda, R., Frueholz, R. P., Kuppermann, A.: J. Chem. Phys. *70,* 1056 (1979)
167. Cuellar, E. A., Marks, T. J.: Inorg. Chem. *20,* 2129 (1981)
168. Bleijenberg, K. C.: Structure and Bonding *42,* 97 (1980)
169. De Hair, J. T. W., Blasse, G.: Chem. Phys. Letters *36,* 111 (1975)
170. Bleijenberg, K. C., De Wit, H. G. M.: J. C. S. Faraday Trans. II *76,* 872 (1980)
171. Van Oosterhout, A. B.: J. Chem. Phys. *67,* 2412 (1977)
172. Valloton, M., Merbach, A. E.: Helv. Chim. Acta *57,* 2345 (1974)
173. Schmitz-DuMont, O., Fendel, H.: Monatsheft *96,* 495 (1965) and *99,* 1285 (1968)
174. De Jong, K. P., Krol, D. M., Blasse, G.: J. Luminescence *20,* 241 (1979)
175. Blasse, G.: J. Electrochem. Soc. *124,* 1280 (1977)
176. Blasse, G.: J. Chem. Phys. *69,* 3124 (1978)
177. Bleijenberg, K. C., Breddels, P. A.: J. Chem. Phys. *72,* 3539 (1980)
178. Kemmler-Sack, S., Seemann, I.: Z. anorg. Chem. *409,* 23 (1974)
179. Ryan, J. L.: J. Inorg. Nucl. Chem. *33,* 153 (1971)
180. Keller, C., Jørgensen, C. K.: Chem. Phys. Letters *32,* 397 (1975)
181. Price, G. R., Ferretti, R. J., Schwartz, S.: Analyt. Chem. *25,* 322 (1953)
182. Runciman, W. A.: Proc. Roy. Soc. (London) *A237,* 39 (1956)
183. Bleijenberg, K. C., Breddels, P. A.: J. Chem. Phys. *72,* 5390 (1980)
184. Lys, J. E. A., Runciman, W. A.: Proc. Phys. Soc. *76,* 158 (1960)
185. Runciman, W. A., Wong, E. Y.: J. Chem. Phys. *71,* 1838 (1979)
186. Parrot, R., Naud, C., Delbecq, C. J., Yuster, P. H.: Phys. Rev. *B15,* 137 (1977)
187. Reisfeld, R.: Structure and Bonding *30,* 65 (1976)
188. Blasse, G.: J. Electrochem. Soc. *115,* 738 (1968)
189. Blasse, G.: J. Inorg. Nucl. Chem. *29,* 1817 (1967)
190. Andrews, L. J., Lempicki, A., McCollum, B. C.: Chem. Phys. Letters *74,* 404 (1980) and J. Chem. Phys. *74,* 5526 (1981)
191. Lempicki, A., Andrews, L. J., McCollum, B. C.: J. Luminescence *25,* 877 (1981)
192. Clarke, R. H., Andrews, L. J., Frank, H. A.: Chem. Phys. Letters *85,* 161 (1982)
193. Weyl, W. A.: Coloured Glasses. London, Dawson's of Pall Mall 1959
194. Gmelins Handbuch der anorganischen Chemie, System-Nummer 55: Uran and Isotope. Weinheim/Bergstr., Verlag Chemie 1936
195. Lam, K. L., Leung, A. F.: J. Non-Cryst. Solids *23,* 385 (1977)
196. Reisfeld, R., Tzehoval, H.: in preparation

197. Ryan, J. L., Keder, W. E.: Advances in Chemistry Series (Washington, D. C.) *71,* 335 (1967)
198. Reisfeld, R., Mack, H., Eisenberg, A., Eckstein, Y.: J. Electrochem. Soc. *122,* 273 (1975)
199. Mack, H., Boulon, G., Reisfeld, R.: J. Luminescence *24,* 111 (1981)
200. Reisfeld, R., Lieblich-Sofer, N., Boehm, L., Barnett, B.: J. Luminescence *12,* 749 (1976)
201. Joshi, J. C., Joshi, B. C., Pandey, N. C., Joshi, J.: J. Non-Cryst. Solids *24,* 361 (1977)
202. Joshi, B. D., Dalvi, A. G. I., Bangia, T. R.: J. Luminescence *10,* 261 (1975)
203. Reisfeld, R.: Multiphonon Relaxation in Glasses in Radiationless Processes (B. Di Bartolo, V. Goldberg, eds.) p. 489. New York, Plenum Press 1980
204. Reisfeld, R., Kalisky, Y.: Chem. Phys. Letters *80,* 178 (1981)
205. Bourcet, J. C., Moine, B., Boulon, G., Reisfeld, R., Kalisky, Y.: Chem. Phys. Letters *61,* 23 (1979)
206. Moine, B., Bourcet, J. C., Boulon, G., Reisfeld, R., Kalisky, Y.: J. Physique *42,* 499 (1981)
207. Powell, R. C., Blasse, G.: Structure and Bonding *42,* 43 (1980)
208. Reisfeld, R., Neuman, S.: Nature *274,* 144 (1978)
209. Reisfeld, R., Kalisky, Y.: Nature *283,* 281 (1980)
210. Reisfeld, R., Greenberg, E., Roth, N., Banks, E., Okamoto, Y.: in preparation
211. Sinha, S. P.: Structure and Bonding *25,* 67 (1976)
212. Brawer, S. A., Weber, M. J.: Phys. Rev. Letters *45,* 460 (1980) and J. Chem. Phys. *75,* 3522 (1981)
213. Brawer, S. A., Weber, M. J.: J. Luminescence *24,* 115 (1981)
214. Jørgensen, C. K.: Mat. fys. Medd. Dan. Vid. Selskab *29,* no. 7 and 11 (1955)
215. Waggener, W. C.: J. Phys. Chem. *62,* 382 (1958)
216. Sjoblom, R., Hindman, J. C.: J. Am. Chem. Soc. *73,* 1744 (1951)
217. Keder, W. E.: J. Inorg. Nucl. Chem. *24,* 561 (1962)
218. Ryan, J. L.: Inorg. Chem. *2,* 348 (1963)
219. Edelstein, N., Brown, D., Whittaker, B.: Inorg. Chem. *13,* 563 (1974)
220. Warren, K. D.: Inorg. Chem. *16,* 2008 (1977) and *19,* 653 (1980) and *20*, 4223 (1981)
221. Selbin, J., Ortego, J. D.: Chem. Rev. *69,* 657 (1969)
222. Ekstrom, A.: Inorg. Chem. *13,* 2237 (1974)
223. Bell, J. T., Friedman, H. A., Billings, M. R.: J. Inorg. Nucl. Chem. *36,* 2563 (1974)
224. Cohen, D.: J. Inorg. Nucl. Chem. *32,* 3525 (1970)
225. Sullivan, J. C., Hindman, J. C., Zielen, A. J.: J. Am. Chem. Soc. *83,* 3373 (1961)
226. Jørgensen, C. K.: Chem. Phys. Letters, submitted
227. Johnston, D. R., Satten, R. A., Schreiber, C. L., Wong, E. Y.: J. Chem. Phys. *44,* 3141 (1966)
228. Cohen, D., Carnall, W. T.: J. Phys. Chem. *64,* 1933 (1960)
229. Gruber, J. B., Morrey, J. R., Carter, D. G.: J. Chem. Phys. *71,* 3982 (1979)
230. Crosswhite, H. M., Crosswhite, H., Carnall, W. T., Paszek, A. P.: J. Chem. Phys. *72,* 5103 (1980)
231. Rabideau, S. W.: J. Phys. Chem. *67,* 2655 (1963)
232. Asprey, L. B., Stephanou, S. E., Penneman, R. A.: J. Am. Chem. Soc. *73,* 5715 (1951)
233. Stephanou, S. E., Nigon, J. P., Penneman, R. A.: J. Chem. Phys. *21,* 42 (1953)
234. Hall, G. R., Markin, T. L.: J. Inorg. Nucl. Chem. *4,* 296 (1957)
235. Coleman, J. S., Keenan, T. K., Jones, L. H., Carnall, W. T., Penneman, R. A.: Inorg. Chem. *2,* 58 (1963)
236. Blasse, G., Dirksen, G. J.: Chem. Phys. Letters *85,* 150 (1982)
237. Flint, C. D., Paulusz, A. G.: Inorg. Chem. *20,* 1768 (1981)
238. Jørgensen, C. K.: J. Inorg. Nucl. Chem. *24,* 1587 (1962)
239. Zink, J. I.: Inorg. Chem. *14,* 555 (1975)
240. Ryan, J. L.: MTP International Review of Science, Inorganic Chemistry, Series 1, *7,* 323 (K. W. Bagnall, ed.) London, Butterworths 1972
241. Jørgensen, C. K.: Mol. Phys. *2,* 96 (1959)
242. Nugent, L. J., Baybarz, R. D., Burnett, J. L., Ryan, J. L.: J. Phys. Chem. *77,* 1528 (1973)
243. Cahn, R. N., Glashow, S. L.: Science *213,* 607 (1981)
244. Flint, C. D., Tanner, P. A.: Mol. Phys. *44,* 411 (1981)
245. Åberg, M.: Acta Chem. Scand. *23,* 791 (1969) and *24,* 2901 (1970)
246. Flint, C. D., Tanner, P. A.: Inorg. Chem. *20,* 4405 (1981)
247. Denning, R. G., Short, I. G., Woodwark, D. R.: Mol. Phys. *39,* 1281 (1980)

248. Fayt, M. M., Vandeput, E. M., De Jaegere, S. A.: Nature *200,* 355 (1963)
249. Perkampus, H. H.: Wechselwirkung von π-Elektronensystemen mit Metallhalogeniden. Berlin – Heidelberg – New York, Springer-Verlag 1973
250. Webb, J. D., Bernstein, E. R.: J. Am. Chem. Soc. *100,* 483 (1978)
251. Miller, J. C., Allison, S. W., Andrews, L.: J. Chem. Phys. *70,* 2524 (1979)
252. Bernstein, E. R., Meredith, G. R., Webb, J.: J. Chem. Phys. *68,* 4066 (1978)
253. Cohen, D.: J. Inorg. Nucl. Chem. *18,* 211 (1961)
254. Giacometti, G., Agostini, G., Gennari, G., Corvaja, C.: J. C. S. Chem. Comm. 1019 (1981)
255. Pyykkö, P., Lohr, L. L.: Inorg. Chem. *20,* 1950 (1981)
256. Tanner, S. P., Vargenas, A. R.: Inorg. Chem. *20,* 4384 (1981)

Author-Index Volumes 1–50

H. Rickert

Electrochemistry of Solids

Introduction

1982. 95 figures, 23 tables. Approx. 260 pages
(Inorganic Chemistry Concepts, Volume 7)
ISBN 3-540-11116-6

Contents: Introduction. - Disorder in Solids. - Examples of Disorder in Solids. - Thermodynamic Quantities of Quasi-Free Electrons and Electron Defects in Semiconductors. - An Example of Electronic Disorder. Electrons and Electron Defects in α-Ag_2S. - Mobility, Diffusion and Partial Conductivity of Ions and Electrons. - Solid Ionic Conductors, Solid Electrolytes and Solid Solution Electrodes. - Galvanic Cells with Solid Electrolytes for Thermodynamic Investigations. - Technical Applications of Solid Electrolytes. Solid-State Ionics. - Solid-State Reactions. - Galvanic Cells with Solid Electrolytes for Kinetic Investigations. - Non-Isothermal Systems. Soret Effect, Transport Processes, and Thermopowers. - Subject Index.

The electrochemistry of solids is of great current interest to research and development. The technical applications include batteries with solid electrolytes, high-temperature fuel cells, sensors for measuring partial pressures or activities, display units and, more recently, the growing field of chemotronic components. The science and technology of solid-state electrolytes is sometimes called solid-state ionics, analogues to the field of solid-state electronics. Only basic knowledge of physical chemistry and thermodynamics is required to read this book with utility. The chapters can be read independently from one another.
The author, well known from his many publications in this area, published a German monograph in 1973 entitles "Einführung in die Elektrochemie fester Stoffe" on this subject which met an enthusiastic response.

Springer-Verlag
Berlin
Heidelberg
New York

MIX
Papier aus verantwortungsvollen Quellen
Paper from responsible sources
FSC® C105338

If you have any concerns about our products,
you can contact us on
ProductSafety@springernature.com

In case Publisher is established outside the EU,
the EU authorized representative is:
Springer Nature Customer Service Center GmbH
Europaplatz 3, 69115 Heidelberg, Germany

Printed by Libri Plureos GmbH
in Hamburg, Germany